RAND

National Security Research Division

Beyond the Nuclear Shadow

A Phased Approach for Improving Nuclear Safety and U.S.–Russian Relations

David E. Mosher
Lowell H. Schwartz
David R. Howell
Lynn E. Davis

Supported by the
Nuclear Threat Initiative

The research described in this report was supported by the Nuclear Threat Initiative. The research was conducted through the International Security and Defense Policy Center (ISDPC) of RAND's National Security Research Division (NSRD).

Library of Congress Cataloging-in-Publication Data

Beyond the nuclear shadow : a phased approach for improving nuclear safety and
    U.S.–Russian relations / David E. Mosher ... [et al.].
       p. cm.
       "MR-1666."
       ISBN 0-8330-3346-8 (pbk.)
       1. Nuclear engineering—United States—Materials—Security measures. 2.
    Nuclear engineering—Russia (Federation)—Materials—Security measures. 3.
    Radioactive substances—Safety measures. 4. International relations. 5. United
    States—Foreign relations—Russia (Federation) 6. Russia (Federation)—Foreign
    relations—United States. I. Mosher, David E.

    TK9152.B43 2003
    621.48'35—dc21

                                                                    2003002127

RAND is a nonprofit institution that helps improve policy and decisionmaking through research and analysis. RAND® is a registered trademark. RAND's publications do not necessarily reflect the opinions or policies of its research sponsors.

Published 2003 by RAND
1700 Main Street, P.O. Box 2138, Santa Monica, CA 90407-2138
1200 South Hayes Street, Arlington, VA 22202-5050
201 North Craig Street, Suite 202, Pittsburgh, PA 15213-1516
RAND URL: http://www.rand.org/
To order RAND documents or to obtain additional information,
contact Distribution Services: Telephone: (310) 451-7002;
Fax: (310) 451-6915; Email: order@rand.org

The past eighteen months have seen remarkable changes in both U.S.-Russian relations and the planned nuclear postures of the two countries. The Bush administration came into office in 2001 with plans to radically remake U.S. relations with Russia in both the political and nuclear arenas, and so far it has made real progress. It has succeeded in negotiating an arms control agreement that promises to cut nuclear forces below today's levels without getting tied up in the usual complications of specified force structures and verification agreements that often take on a life of their own. The very existence of this agreement suggests that relations have improved between the United States and Russia and, perhaps more important, that the Cold War's long shadow of nuclear confrontation is beginning to recede.

Despite those changes, however, the risk of accidental and unauthorized use of nuclear weapons remains. Both countries continue to keep their nuclear forces ready to launch within minutes, the only explanation for which is that they continue, either by habit or design, to view each other in nuclear terms and to posture themselves to respond to attack by the other. Russia's economic difficulties over the past decade have made the situation worse: Russia's early-warning system is in tatters, and its military has suffered both morale and discipline problems.

The report explores a wide range of approaches by which the United States and Russia could improve nuclear safety. The authors argue that U.S.-Russian relations and nuclear safety are integrally linked and should not be viewed in isolation—improvements in one will lead to improvements in the other. Consequently, they recommend that the two countries take a phased approach to the problem, one that starts with immediate U.S. unilateral actions and commitments designed to demonstrate that the United States is serious about improving relations and nuclear safety. The authors also suggest specific actions that can be taken in the near and medium term as U.S.-Russian relations improve. The long-term goal of this process is to eliminate the nuclear element of the relationship altogether.

This research was made possible by a grant from the Nuclear Threat Initiative. The report is intended for those concerned with nuclear issues, arms control, or U.S.-Russian relations more broadly.

The study was conducted within the International Security and Defense Policy Center (ISDPC) of RAND's National Security Research Division (NSRD). NSRD conducts research and analysis for the Office of the Secretary of Defense, the Joint Staff, the unified commands, the defense agencies, the Department of the Navy, the U.S. intelligence community, allied foreign governments, and foundations.

CONTENTS

FIGURES

TABLES

SUMMARY

The last decade has brought significant changes in the relations between the United States and Russia. At the political level, these changes have most recently been demonstrated, in extraordinary fashion, by Russia's providing active assistance in the war on terrorism, even helping the United States establish basing rights in Central Asia. Changes at the nuclear level have also been notable, as evidenced by the May 2002 signing of the Moscow Treaty statements in which Presidents George W. Bush and Vladimir Putin each agreed to reduce long-range nuclear forces to 2,200–1,700 over the next 10 years, down from more than 10,000 each in 1990.

Many of the nuclear dangers that characterized the Cold War—a surprise nuclear attack or a crisis in Europe or Asia that could lead to nuclear war—have receded. Now that there is no longer an ideological conflict as motivation or armies poised in Central Europe to spark a crisis, neither country views nuclear war with the other as likely.

Yet despite the steps taken by both countries to put Cold War hostilities behind them, an important nuclear risk remains—specifically, that of accidental and unauthorized use of nuclear weapons. This risk persists for three reasons. First, although both countries have significantly reduced their nuclear forces, they still retain nuclear postures and deterrence doctrines formulated when tension between them was much higher than it is today. Inherent in these nuclear postures, which are based on rapid delivery of a massive nuclear retaliatory strike, are concerns about the potential for an accidental or unauthorized launch.

Second, Russia's economic and social troubles have created a new set of problems that contribute to the continuing risk of accidental and unauthorized use. Russia's resource shortages have caused its survivable nuclear forces to plummet in both size and readiness. Its fleet of ballistic missile submarines has been decimated; most of them are decommissioned or rusting in port, and only one or two are at sea at any time. Few of its mobile missiles are deployed in the field, and many of its intercontinental ballistic missiles (ICBMs) are well beyond their planned service lives. In a severe crisis, these vulnerabilities may push Russia toward a strategy of launching its forces quickly, at the first signs of attack, to ensure their survival—a posture leaving little time for decisions and possibly leading to accidental use of nuclear weapons. This state of affairs is further complicated by an early-warning system in serious disrepair and by Russia's increasing reliance on nuclear weapons to compensate for its atrophying conventional forces. Moreover, the risk of unauthorized launch has been heightened by personnel reliability problems arising from the social and economic upheavals Russia has experienced over the past decade, as well as by endemic problems with organized crime and its ties to separatist groups.

Third, Russia's vulnerabilities are accentuated by the design of U.S. forces, which were built to destroy Russia's silo-based missiles. The Trident submarine, with its accurate missiles and powerful warheads, has allowed the United States to make a significant portion of those Russian forces vulnerable. As long as Russia could deploy survivable ballistic missile submarines and road-mobile and rail-based ICBMs—which it could in the 1980s—it ensured that enough of its forces would survive to retaliate against a U.S. strike. Now, however, with only a few of Russia's survivable forces able to leave their bases, the United States is closer to being able to destroy Russia's forces than ever before. This situation may make Russia feel even more vulnerable in a crisis and may heighten its incentives to launch its nuclear forces quickly if it fears a nuclear attack.

Clearly, the improved political climate between the United States and Russia somewhat mitigates this risk of nuclear use, making it less dangerous than it might have been during the Cold War. Yet the improved environment also highlights the fact that both countries' nuclear postures do not reflect post–Cold War realities. The world is rapidly approaching a time when any risk of accidental or unauthorized nuclear use that does remain will be considered an unacceptable anachronism.

Our study focused on today's remaining risk of accidental and unauthorized use of U.S. and Russian nuclear weapons, examining in detail a number of steps the United States and Russia could take (both unilaterally and cooperatively) to reduce the risk and to bring their nuclear postures more in line with current political realities.[1] We found that while several promising steps could be taken, no single one will eliminate the risk that lingers. Furthermore, since some of the steps require the upending of years of orthodoxy about deterrence and the way that nuclear forces are postured and operated, we concluded that a phased approach is likely to be the most productive way to improve nuclear safety. Perhaps our most important finding, however, is that nuclear safety and U.S.-Russian relations are inextricably linked. Nothing will do more to improve nuclear safety than improved U.S. relations with Russia, and because the nuclear dimension still looms large in the U.S.-Russian relationship, steps to improve nuclear safety can lead to a relationship defined less and less by nuclear weapons. And because of the challenges of developing and implementing actions that will improve nuclear safety and U.S.-Russian relations, we believe success requires strong Presidential leadership and commitment.

Scenarios for Nuclear Use

To determine the underlying causes of possible accidental or unauthorized nuclear use, we carefully examined the types of scenarios in which such use could occur. Our analysis suggests that there are three basic types, as shown in Figure S.1. The first is an intentional unauthorized launch. Such scenarios, brought about by a terrorist or a rogue commander (a commander who takes control of the nuclear forces he commands), have always been a concern, and both the

---

[1]Another set of nuclear risks has been created by the Soviet Union's dissolution and Russia's economic difficulties: the danger that nuclear weapons, materials, and know-how could spread to countries that want to develop nuclear weapons. This problem is very serious and could have long-term implications for U.S. security, but it was beyond the scope of our analysis. We did, however, keep it in mind, designing the options we examined to ensure they did not exacerbate the problem of proliferation from Russia and, perhaps, might even improve the situation.

Type I: Unauthorized Launch
- Intentional launch by rogue commander or terrorist

Type II: Launch by Mistake
- Training accident
- System malfunction

Type III: Intentional Launch Based on Incorrect Information
- Malfunction of early-warning system
- Incorrect interpretation of nonthreatening event
- Misperception of nuclear attack by third country or terrorists
- Misperception of accidental nuclear detonation on own territory
- Misinterpretation of simulated training attack as real attack

**Figure S.1—Possible Scenarios for Accidental or Unauthorized
Nuclear Use, by Type**

United States and Russia went to extraordinary lengths during the Cold War to ensure they did not happen. The breakdown of order in Russia, the economic difficulties and low morale of its military personnel, and the rise in organized crime and separatist violence have increased concern about the security of its nuclear forces.

In the second type of scenario, a missile is launched by mistake. In this case, the country that launches the missile has no intention of doing so; the launching occurs through a malfunction in a weapons system or during a training accident. In the past, both Russia and the United States made great efforts to guard against such accidents, and most, if not all, of the safeguard systems and procedures that were put into place remain in place today. The United States has even enhanced the safeguards in recent years for its ballistic missile submarines. Nevertheless, the probability of a mistaken launch has never been zero, and the economic and social problems in Russia have heightened concerns in the West about this problem.

In the third type of scenario, nuclear weapons are launched intentionally but based on incorrect or incomplete information. Such a scenario can occur in different ways. For example, early-warning systems might malfunction, indicating that an attack was under way when in fact it was not; or a nonthreatening event might be misinterpreted as an attack. These two types of events happened during the Cold War to the United States and Russia (although, in each case, the error or malfunction was discovered before weapons were launched), and each country developed a two-tiered early-warning system (radar on land and infrared sensors in space) in part to guard against such events. But Russia's space-based system is now essentially out of order, leaving Russia with only one type of early-warning system and greatly increasing the chance that an erroneous indication of attack could be mistaken as real. Similarly, lacking high-quality early-warning information, Russia could interpret a nuclear accident or an attack by a third country as an attack by the United States. The chances of this type of event occurring, however, are somewhat mitigated by the overall positive relations between the United States and Russia. Because the likelihood of nuclear conflict is much lower than it was during the Cold War, both nations' leaders are far less likely to believe a nuclear attack has been launched against them than they were during the tense periods of the 1960s and 1970s.

Factors Contributing to Nuclear Use

Using our set of possible scenarios (see Figure S.1), we tried to determine the underlying factors that might cause any one of them to occur. We identified seven such factors, which are listed here in no particular order:

- Nuclear forces kept at high day-to-day launch readiness

- Perceived vulnerability of nuclear forces or command and control systems

- Inadequate early-warning information

- Short decision times

- Deterrence doctrine or posture reliant on launch on warning or launch under attack

- Inadequate security and control of nuclear forces and weapons

- Inadequate training precautions

In our estimation, nuclear use is likely to result only from a combination of these factors—e.g., a launch-on-warning doctrine plus inadequate early-warning information—rather than from any single one. We mapped these contributing factors to each type of scenario for accidental or unauthorized use of nuclear weapons before we began devising measures that might reduce the risk of such use occurring, which was the primary focus of our analysis.

Possible Options for Reducing Risk

A wide variety of solutions have been proposed as ways to reduce the risk of accidental or unauthorized use of nuclear weapons. Some are relatively straightforward extensions of current policies; others require fundamentally new approaches to how both the United States and Russia operate and think about nuclear weapons.

We considered and analyzed many possible actions that attempt to address one or more of the underlying factors for accidental or unauthorized nuclear use. They can be distilled into 16 general approaches, each of which can be implemented in many ways. From the list of 16, we chose for detailed analysis those that we believe show promise or that have been proposed by others. We arrived at the following 10 approaches, or options:

1. Provide assistance for improving Russia's early-warning radars or satellites.

2. Establish a joint, redundant early-warning system by placing sensors outside U.S. and Russian missile silos.

3. Immediately stand down all nuclear forces to be eliminated under the 2002 Moscow Treaty.

4. Pull U.S. strategic ballistic missile submarines away from Russia.

5. Keep U.S. attack submarines away from Russia.

6. Remove W-88 warheads from Trident missiles.

7. Reduce day-to-day launch readiness of 150 ICBMs in silos.

8. Reduce day-to-day launch readiness of all nuclear forces.

9. Install destruct-after-launch (DAL) mechanisms on ballistic missiles.

10. Deploy limited U.S. missile defenses.

An additional advantage of these 10 options is that each one is a possible solution for more than one of the contributing factors.

We assessed each option against a set of criteria reflecting the degree to which it could contribute to nuclear safety; its effect on current U.S. strategies, targeting plans, and stability; and the degree to which it could be implemented easily, monitored reliably, advance or hinder U.S. nonproliferation objectives, and improve U.S.-Russian relations.

Our goal was to be as specific as possible about the measures that would be taken in each option, the reason being simply that the details of an option's construction often determine how effective and successful the option can be. For example, placing sensors outside U.S. missile silos to detect launches would improve Russia's access to reliable early-warning information. It would add 20 minutes or so to decision time, extra time important for determining whether an attack is real. Perhaps most important, it would tell Russia that an attack is not under way by showing the status of each individual silo, something that space-based systems could not accomplish. Yet such a system is worse than nothing at all if it is not highly reliable and designed to be resistant to generating false alarms every time a mouse chews a cable, a thunderstorm passes through, or the power goes out. A system that issues frequent false alarms will be ignored; a system that is vulnerable to large systemic errors can lead to an accidental use of nuclear weapons—the very problem the option is intended to avoid. So we specified as carefully as we could how such a system might be designed to avoid these problems—for instance, by having backup power sources, using redundant communication modes, and using different sensors that measure different physical phenomena.

General Conclusions

Our analysis led to five general conclusions about improving nuclear safety:

- The risk of accidental or unauthorized nuclear use is too high given the markedly improved relationship between the United

States and Russia. This is in part because nuclear weapons now play a role out of proportion to other aspects of the relationship.

- Although several options we examined show promise for reducing the risk of accidental or unauthorized use of nuclear weapons, no single approach will eliminate the risk by itself. In fact, even a combination of fairly radical options will not entirely eliminate the two fundamental issues giving rise to that risk: the asymmetries in U.S. and Russian force postures and the fact that each country continues to believe it needs a credible deterrent against the other. Nevertheless, both nations can take important steps to minimize the nuclear use risk that remains today.

- Nuclear safety and U.S.-Russian relations are closely intertwined. The more each country views the other as a threat, the more difficult it will be to reduce the risk of accidental or unauthorized use of nuclear weapons. Conversely, the better the relations are, the more likely that nuclear weapons will recede as an important factor in the relationship and the easier it will be to take significant steps to improve nuclear safety.

- A successful strategy for limiting nuclear dangers should, therefore, seek operational changes in the U.S. and Russian nuclear postures, as well as improvements in the trust and cooperation between the two nations. This should be a mutually reinforcing process in which near-term improvements in nuclear safety build confidence and trust between Russia and the United States, thereby enabling more-extensive steps in the medium and long term.

- Direct Presidential involvement is required to adequately address the nuclear use risk. The issues are too fundamental, affect too many policies, and cross too many departments, services, and agencies for an uncoordinated approach to succeed. The effort must start with strong Presidential leadership and commitment. The goals should be to improve nuclear safety, to improve U.S.-Russian relations, and to preserve a survivable deterrent. Essential to this effort is close military-to-military cooperation between the two nations, both for devising options that make operational sense and for implementing them.

Specific Recommendations: A Phased Approach for Improving Nuclear Safety

To address the lingering risk of accidental and unauthorized use of nuclear weapons, we recommend that the United States and Russia establish a "Nuclear Safety Initiative." And because the task of improving nuclear safety is likely to be a challenging and dynamic process, we recommend that the initiative use a phased approach, as illustrated in Figure S.2. The objective of the first, immediate steps (six months to one year) is to improve nuclear safety while simultaneously improving the climate between the two countries. If these steps are successful, they will create opportunities for more involved steps in the near term (the next two to three years). Success in the near term will then make other, more-difficult steps possible in the medium term (five to seven years). The timeline illustrated here is intended to reflect what we think may be possible from today's vantage point. The timing could be quicker if conditions and leadership allow.

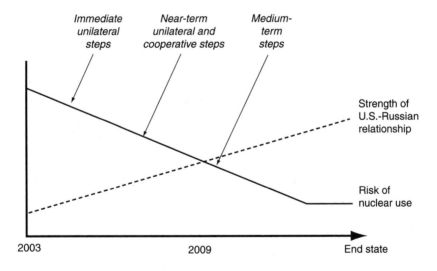

Figure S.2—Phased Approach for Improving Nuclear Safety
and U.S.-Russian Relations

A central reason for the phased approach is that some options for improving safety would push too far beyond current deterrence practices and orthodoxies to be acceptable. Since changes in thinking and nuclear culture will be required, these options are more likely to be successful once success is achieved in other areas and the importance of nuclear weapons in the U.S.-Russian relationship has receded somewhat. As a result, immediate steps to improve the climate between the two countries should be pursued even if they will have only a small effect on safety.

Figure S.3 lists the immediate, near-term, and medium-term steps we believe could start the process of improving nuclear safety (and U.S.-Russian relations). As shown, the United States could immediately take several steps: standing down all forces to be deactivated under the Moscow Treaty, pulling ballistic missile submarines and attack submarines away from Russia, and reducing the launch readiness of one-third of its ICBMs. These actions could be implemented within six months to one year. Also shown are several steps the United States could commit to at this time. These steps, aimed at helping Russia improve its access to reliable, high-quality early-warning information, would take more time to implement, in part because they would require consultations.

The hope is that Russia will respond with its own unilateral measures, standing down the forces that it will eliminate under the Moscow Treaty, keeping its submarines away from U.S. coasts, and committing to cooperate with the United States on improving access to early-warning information. However, the unilateral U.S. actions during this first phase should not presume or depend on Russian reciprocation. The goal of these steps is to reduce the risk of nuclear use and to demonstrate that nuclear weapons are becoming less important in U.S. relations with Russia.

In the near term, the United States could eliminate the forces it had stood down during the first phase (as could Russia if it had followed suit) and begin implementing steps to improve early-warning information. The United States and Russia could also begin consultations on more difficult issues, including steps to reduce the launch readiness of nuclear forces and to install destruct-after-launch (DAL) mechanisms on ballistic missiles. In the medium term, possible steps include installing sensors on ICBM silos to monitor reduced launch

6 months to 1 year	2 to 3 years	5 to 7 years
Immediate Unilateral Steps	*Near-Term Unilateral Steps*	*Medium-Term Steps*
• Stand down U.S. forces to Moscow Treaty levels	• Eliminate Moscow Treaty forces	• Take equal number of silo-based ICBMs off alert
• Pull SSBNs away from Russia	• Put EW sensors on U.S. silos	• Install sensors on silos to monitor reductions in launch readiness
• Pull U.S. attack subs back	• Remove W-88 warheads	
• Reduce launch readiness of 1/3 of U.S. silo-based missiles	*Begin Consultations on:*	• Adopt a new deterrence strategy
		• Deploy limited missile defense
Commit to:	• Further improving Russian EW systems	• Deploy DAL system (mid-course)
• Put EW sensors on U.S. silos	• Installing DAL systems	• Continue negotiations on further steps to reduce launch readiness
• Fund Russian EW radar	• Reducing launch readiness	
• Continue RAMOS program		

| 2003 | 2006 | 2010 |

NOTE: SSBN = ballistic missile submarine; EW = early warning; DAL = destruct after launch; RAMOS = Russian-American Observational Satellite.

Figure S.3—Potential Steps for Improving Nuclear Safety

readiness and adopting a new deterrence strategy, one that is less dependent on quick responses to attacks. In addition, steps to reduce the effects of nuclear use if it did occur—i.e., limited missile defenses and DAL systems—could be taken if the appropriate technology is ready.

Of course, the exact steps that the United States and Russia take could be different from those we are suggesting, particularly in the medium term and beyond, as both sides determine the best ways to improve nuclear safety and further reduce the importance of the nuclear dimension in their relationship. The details are secondary; what is most important is that the process begin immediately.

The phased approach to our recommended Nuclear Safety Initiative is based on the premise that nuclear safety, U.S.-Russian relations, and U.S. security more broadly are inextricably linked. Progress in one area can improve the situation in another. Our approach represents the best path for addressing the risk of nuclear use and at the same time allowing Russia and the United States to maintain nuclear forces that are sized and postured appropriately for each stage of their improving relations. Ultimately, however, the nuclear safety problem can be fully addressed only when nuclear weapons are no longer a factor in U.S.-Russian relations, much like they are not a factor in relations between Britain and France.

Given the improving relations between the United States and Russia and the emerging U.S. security context, the current moment offers a historic opportunity to address one of the more vexing problems left from the Cold War: how to reduce the risk of accidental or unauthorized nuclear use to as close to zero as possible. Only a sustained, coordinated effort can solve this problem, and such an effort must start with Presidential leadership and commitment.

ACKNOWLEDGMENTS

This research would not have been possible without the financial support of the Nuclear Threat Initiative or the keen personal interest in the project by its Co-Chairman and Chief Executive Officer, Senator Sam Nunn.

Many people helped us during our research, which for one of us stretches back more than four years. Geoffrey Forden and Ted Postol of MIT's Center for Strategic Studies shared their analysis and assessments of the state of Russia's early-warning systems. Pavel Podvig, Evgeny Miasnikov, Timur Kadyshev, and Anatoli Diakov of the Center for Arms Control, Energy, and Environment at the Moscow Institute of Physics and Technology shared their technical knowledge and views of Russia's nuclear forces and postures. Vladimir Sukharuchken and academicians Evgeny Velikhov and Nicholai Panomarev-Stepnoi of the Kurchatov Institute of the Russian Academy of Sciences provided intriguing concepts for monitoring systems. Valery Yarynich and Petr Romashkin of the Russian State Duma shared their views on launch readiness reduction. Frank von Hippel provided useful comments, suggestions, and inspiration along the way. Many experts from the U.S. government, Navy, Air Force, and Department of Energy weapons laboratories aided our research, listening to our ideas and proposals and providing technical insights.

We would also like to thank Bruce Blair, General Eugene Habiger, Arnold Kantor, Jan Lodal, Joan Rohlfing, Charles Curtis, and Ted Warner for their comments on the manuscript and insightful discussion of key issues. We are also indebted to Michael Lostumbo and

David Ochmanek of RAND, and Robert Nurick and Alexei Pikayev of the Carnegie Center in Moscow for their suggestions during the project's formative period. Ted Warner was instrumental in getting the project started and provided intellectual guidance and the benefits of his experience throughout the process. Roger Molander of RAND and Janne Nolan of Georgetown University deserve special thanks for their careful reviews of the draft and their thoughtful comments. Mickeeta Brooks deserves many thanks for her patient preparation of the draft manuscript. We would also like to thank Jeri O'Donnell, who patiently edited the manuscript, Janet DeLand, who prepared it for publication, and Christopher Kelly, who oversaw the publications process.

The contributions of all these individuals greatly improved the quality of our study. However, we alone bear responsibility for the final product.

ABBREVIATIONS

ABM	anti-ballistic missile
BMDO	Ballistic Missile Defense Organization
CBO	Congressional Budget Office
DAL	destruct after launch
DALcc	DAL control center
DoD	Department of Defense
DSP	Defense Support Program
GPALS	Global Protection Against Limited Strikes
GPS	global positioning system
ICBM	intercontinental ballistic missile
IFT	integrated flight test
LPAR	large phased-array radar
NASA	National Aeronautics and Space Administration
NSC	National Security Council
PD	Presidential Directive
RAMOS	Russian-American Observational Satellite
SBIRS	Space-Based Infrared System
SDI	Strategic Defense Initiative
SIOP	Single Integrated Operational Plan

SLBM	submarine-launched ballistic missile
SORT	Strategic Offensive Reductions Treaty
SSBN	ballistic missile submarine
START	Strategic Arms Reduction Treaty

Chapter One

BACKGROUND AND MOTIVATION FOR IMPROVING NUCLEAR SAFETY

Since the dawn of the missile age, national security decisionmakers have sought to achieve a delicate balance between having a credible nuclear deterrent and ensuring the safety of nuclear weapons from unauthorized or accidental use. During the Cold War, when the United States and Russia perceived each other as a serious threat, the balance was weighted heavily toward the credibility of the deterrent, which was seen in terms of the ability to mount a massive and immediate nuclear retaliatory strike. Nevertheless, both countries took strong steps to reduce the risk of accidental or unauthorized use to the extent their nuclear postures would allow. Today, with the Cold War over, the ideological sources of the superpower conflict gone, and growing concern about the proliferation of weapons of mass destruction, a shift in the balance toward nuclear safety seems appropriate. The attacks of September 11, 2001, have underscored the need for this shift by pointing out the nature of the new nuclear threats that both the United States and Russia are likely to face. Moreover, Russia's reaction to the attacks and its support for the war on terrorism have demonstrated both the broad improvements that have been made in the relationship between the two former Cold War antagonists and the feasibility of pursuing such a shift.

This report provides a roadmap for how this shift toward nuclear safety could take place. First, it examines the types of scenarios that might lead to unauthorized or accidental use of nuclear weapons. It then considers contributing factors (such as nuclear forces being kept on high alert and short decision times) that might lead to unauthorized or accidental use. Finally, it develops a set of options for dealing with each contributing factor and weaves these options together to provide a phased approach for improving nuclear safety.

CONCERNS ABOUT THE CURRENT SITUATION

Today's concerns about nuclear safety are driven by several factors, including the nature of nuclear forces in both Russia and the United States, the status of Russia's early-warning system, Russia's economic difficulties, and several recent geopolitical trends.

Historical Asymmetries in Nuclear Forces

The characteristics of the U.S. and Russian nuclear forces, as they have evolved historically, have contributed to concerns about safety. Russia has traditionally been a land-based power, its nuclear forces heavily emphasizing intercontinental ballistic missiles (ICBMs) based in fixed silos hardened to withstand the effects of nuclear blasts. These missiles are reportedly ready to launch within a few minutes.

During the 1980s, the U.S. ICBM force was modernized to make it more accurate and survivable. The more significant development during that period, however, was the deployment of the Trident submarine, which represented a quantum leap in capability over previous generations of submarines. Its D5 missiles were as powerful and accurate as the best ICBMs, and it carried the W-88—the U.S. arsenal's most powerful warhead—specifically designed to attack and destroy hardened Russian silos. But the Trident also gave the United States something else: the ability to attack those hardened targets quickly. Trident missiles can reach their targets in 10 to 15 minutes if they are launched close to Russia; an ICBM would take at least 30 minutes. The Trident's combination of accuracy, lethality, and speed gave the United States the ability to deliver not only a retaliatory strike against Russian nuclear forces, but also a devastating first strike.

Russia responded to the increased accuracy of U.S. missiles during the 1980s by boosting the survivability of its ICBM force: it deployed some of its missiles on railcars and off-road trucks. When these mobile missiles were dispersed, they became almost as survivable as submarines. (When clustered together in their garrisons, however, they are far more vulnerable than silo-based ICBMs.) Russia also had a sizable nuclear submarine fleet, a portion of which remained at sea and thus was highly survivable. Hence, despite U.S. *technological* advances during the 1980s, both countries retained large and highly survivable nuclear forces.

Russia's Declining Nuclear Forces

The nuclear balance that had been established was fundamentally altered during the 1990s. While Russian forces deteriorated during and after the Cold War, making them increasingly vulnerable to a first strike, the United States retained much of its counterforce capabilities and deployed more Trident submarines with D5 missiles. This situation is likely to become more pronounced as Russian nuclear forces continue to decline in the first decade of the 21st century.

Today, Russia keeps all but a few of its mobile missiles in garrison, where they can be easily destroyed. Only one or two regiments (nine to 18 missiles) of its 360 road-mobile missiles are dispersed in the field at any one time. Its rail-based missiles are restricted to garrison by an order President Yeltsin issued in 1994. To boost the survivability of the mobile missiles it has in garrison, Russia could launch them through doors in the roofs of their garages before U.S. missiles arrive. As for Russia's ballistic missile submarines, only a small fraction of them (perhaps one or two) are kept at sea. The rest are in port, where they are very vulnerable to attack—one nuclear warhead can destroy most of the submarines at a base. Probably to improve their survivability, Russia's modern submarines are capable of quickly launching missiles from pier side that can hit targets in the United States. The final piece of Russia's nuclear triad, its bomber force, has always been kept at relatively low levels of readiness and is rarely used today. In the event of a significant surprise attack, few bombers are likely to survive.

By contrast, the United States retains a large and survivable nuclear force divided roughly equally among ICBMs, bombers, and submarines. Like Russia, the United States keeps its ICBMs (all based in silos) at high levels of alert, ready to launch within a few minutes. During the Cold War, the United States also kept its bombers on high alert, a portion of them either airborne or ready to take to the air within a few minutes. But since 1991, U.S. bombers have been taken off alert. Although roughly half of them still retain their nuclear mission, they spend much of their time training for or participating in nonnuclear conflicts. U.S. ballistic missile submarines provide the most survivable force for the United States and cause the most discomfort for Russia. Unlike Russia, the United States keeps a large portion of its submarines—roughly 60 percent—at sea, even today. This provides a large survivable force capable of delivering at least 1,000 warheads to targets in Russia.

In sum, whereas the United States has a large, survivable nuclear force with some 1,300 warheads deployed at sea, Russia has very few nuclear forces that could survive a surprise U.S. attack—only about 20 to 200 warheads—if it rode out the attack before launching a retaliatory strike. Although we do not know for certain, Russia may regard this number of nuclear forces as insufficient to deter the United States in a crisis, and may therefore be relying on a launch-on-warning strategy—a standard approach for maximizing the size of a retaliatory attack—that would allow it to retaliate with some 3,000 warheads. The launch-on-warning approach to nuclear warfare is, however, very destabilizing, because its proper execution requires an extremely rapid reaction—probably within 10 or 15 minutes. This means there is very little time to verify that early-warning information from satellites and land-based radars is correct.

Russia's Internal Problems

Three distinct internal problems, associated with Russia's economic collapse during the 1990s, exacerbate the concerns inherent in a nuclear strategy based on retaining forces on high alert and launching on warning:

1. Russia's social and economic problems have caused a substantial decline in its conventional force capabilities. Russia perceives its conventional forces as no match for the modern high-tech forces the West has demonstrated in the Persian Gulf, Kosovo, and Afghanistan. As a result, Russia increasingly relies on nuclear weapons to counter the West's conventional superiority and to deter its southern neighbors.

2. Russia's early-warning system has deteriorated significantly. Like the United States, Russia relies on a combination of satellite-based infrared sensors and ground-based radars to provide early warning of an attack and to reduce the chances of mistakes. Satellites provide the earliest warning of an attack; they detect the hot exhaust from missiles as they streak into space. Radars track the missiles as they get closer to the target. The problem is that both Russia's satellite and radar networks have holes in them today. Analyses by the Congressional Budget Office (CBO) and others have shown that Russian satellites currently have little, if any, ability to detect missiles launched from U.S. Trident submarines.[1] The satellite network observing U.S. ICBM fields has only one or two of its six satellites working today, which provides coverage for only about six hours a day. As for Russia's radar network, it, too, is incomplete: It has a large gap to the east and a smaller gap to the west through which Trident missiles could fly all the way to Moscow without being seen. The implications of Russia's blindness to nuclear attack are extremely troubling when combined with the compressed decision time required to execute a launch-on-warning strategy.

3. The general disorder in Russia today creates much uncertainty about the security and control of nuclear forces and materials. The far-flung deployments of Russia's nuclear forces and materials, the existence of separatist and terrorist groups, and the strong presence of organized crime in Russia combine to make this situation particularly dangerous.

U.S. Contributions to Nuclear Risk

The risk of accidental or unauthorized nuclear use is not created by Russia alone. The United States exacerbates the risk by continuing to posture its nuclear forces in a manner suitable to a nuclear damage limitation strategy—i.e., able to destroy a large portion of Russia's

---

[1]See Geoffrey Forden, "Letter to the Honorable Tom Daschle Regarding Improving Russia's Access to Early-Warning Information" (Washington, DC: Congressional Budget Office), September 3, 1998, pp. 1–14; Geoffrey Forden, *Reducing a Common Danger: Improving Russia's Early Warning System* (Washington, DC: CATO Institute), May 3, 2001; and Pavel Podvig (ed.), *Russian Strategic Nuclear Forces* (Cambridge, MA: MIT Press), 2001, pp. 428–432.

nuclear forces before they can retaliate against the United States. This strategy involves having a large number of counterforce weapons deployed at the ready—in other words, ready to be launched within a few minutes or perhaps hours—in order to rapidly destroy a high percentage of Russia's nuclear forces. Similarly, the United States continues to patrol its attack submarines near the home bases and operating areas of Russia's increasingly vulnerable ballistic missile submarine force, where they can track the few of these submarines that Russia manages to put at sea. Furthermore, U.S. conventional forces in Iraq, Yugoslavia, and Afghanistan have demonstrated the ability to destroy hardened targets with nonnuclear, precision-guided weapons. Many Russian analysts are concerned that such weapons could be used against Russian nuclear targets. On the diplomatic and political side, the United States has shown a willingness to build a large national missile defense system even if doing so means abandoning the Antiballistic Missile Treaty. Russians remain concerned that a future U.S. national missile defense system, along with a large number of U.S. counterforce weapons (both nuclear and conventional), could severely limit, if not eliminate, Russia's nuclear deterrent.

OPPORTUNITIES FOR IMPROVING NUCLEAR SAFETY IN AN ERA OF IMPROVING RELATIONS

Fortunately, the end of the Cold War and the corresponding improvement in U.S.-Russian relations have created an opportunity for both countries to take steps to improve nuclear safety. Some steps in this direction have already been taken, such as sharp reductions in forces and the sharing of early-warning information, and further improvements in relations will make other measures possible as well. Moreover, steps taken to improve nuclear security can improve U.S.-Russian relations by reducing the relevance of nuclear weapons to the relationship.

The improvements in U.S.-Russian relations that have taken place so far are clearly indicated by both nations' reactions to the attacks of September 11, 2001. For the first time since the Second World War, the United States and Russia find themselves allied against a common foe. Russia's role in the war on terrorism has been substantial: encouraging the use of former Soviet bases in Central Asia, arming and funding Afghani groups opposed to the Taliban regime, and actively supporting U.S. initiatives in the international community. For the first time, it is possible to see a future in which Russia is a full-fledged member of the Euro-Atlantic community.

This new geostrategic environment poses difficult questions for U.S. and Russian strategists about deterrence requirements in the future and the appropriate size, posture, command and control infrastructure, and strategy for each nation's nuclear arsenal.

During the Cold War, the primary U.S. security goal was to deter a Soviet/Warsaw Pact invasion of Europe and a nuclear strike by the Soviet Union. The United States pursued this goal by building and deploying tens of thousands of nuclear weapons. Today, the greatest threat from Russia comes not from its strength but from its weakness. Russia's dysfunctional economy and eroded security systems have undercut its control of the vast stockpile of weapons, materials, and know-how accumulated during the Cold War, thereby increasing the risk that they could flow to terrorist groups or other hostile forces.

As Russia and the United States begin to explore the form and structure of their new deterrence postures, several features are likely to carry over from their Cold War postures.[2] First, both nations are

---

[2]We have attempted to avoid prejudging what nuclear strategy and posture are appropriate for the new strategic era. While not taking a stance on nuclear strategy issues, we do highlight nuclear safety options that *do* and *do not* require a change in nuclear strategy and posture. It is also important to point out that many steps can be taken to improve nuclear safety in the absence of a wholesale change in U.S. nuclear strategy.

However, we feel it important to say that the long-running U.S. nuclear strategy of damage limitation (i.e., the requirement to destroy as many Russian nuclear weapons as possible before they can be used to retaliate against the United States) would be a serious impediment to improving nuclear safety. A damage limitation strategy necessitates a nuclear posture that includes a large number of counterforce weapons ready to launch within minutes to rapidly destroy Russia's nuclear forces. For example, a Trident submarine must be within a certain distance of its designated targets with its missiles prepared to fire within minutes if it is to destroy Russian ICBM forces before they can be launched. This posture is directly at odds with attempts to assure Russia it can move to a more relaxed nuclear posture and thereby decrease the risk of an unauthorized or accidental nuclear launch.

Whether a damage limitation nuclear strategy is necessary today to deter Russia from attacking the United States is a point of serious contention among nuclear strategists both within and outside the U.S. government. Some strategists contend that

likely to retain some kind of retaliatory deterrent. Although the new strategic environment is vastly different from that of the Cold War, nuclear weapons remain the ultimate deterrent against nuclear attacks against the homeland or important regional allies. Second, the historical asymmetries in Russian and U.S. nuclear forces and operating practices are likely to persist for the foreseeable future.

Despite the likely persistence of these Cold War features, however, new factors are creating the potential for a very different and greatly improved nuclear relationship. The most important of these new factors is the improving and increasingly cooperative nature of U.S.-Russian relations. Ten years after the Cold War, it seems highly anachronistic that both nations retain thousands of nuclear weapons on high alert and tolerate the associated risk of unauthorized and accidental nuclear use. Another important factor is the changing nature of the strategic threat confronting the United States as a result of the September 11 terrorist attacks. Although the current crisis does not represent the full spectrum of strategic issues the United States is likely to face during the 21st century, it will probably cause strategic thinkers to reevaluate the role of nuclear weapons in this new geo-strategic era.

At the same time, the obstacles to reducing the role of nuclear weapons in the U.S.-Russian relationship remain formidable. Despite recent cuts in their nuclear arsenals, both countries retain very large numbers of nuclear forces. Even if all reductions recently agreed to in

---

a more flexible nuclear doctrine is more appropriate today, one that emphasizes countervalue attacks in the event of hostility with another nuclear-armed state, which could be Russia, China, or a rogue state. These analysts argue that a damage limitation strategy is inappropriate because of the vastly improved relations between the United States and Russia and because many features that led to the introduction of a counterforce strategy in the first place disappeared with the end of the Cold War. For example, the possibility of conventional war between Russia and the United States is now regarded as such a remote possibility that the Pentagon has completely removed it as a planning scenario for sizing and modernizing U.S. conventional forces. The analysts suggest that nuclear planning should be similarly altered, directed away from specific scenarios focused on Russia and toward more general nuclear scenarios. Public statements by officials in the Bush administration suggest that the 2001 Nuclear Posture Review took steps in this direction.

We do not know at this time whether the United States still retains a damage limitation strategy. However, U.S. forces have been designed and operated for such a strategy, and demonstrating a retreat from that strategy will be difficult in the absence of overt changes to the postures or the forces themselves.

the Moscow Treaty are achieved by 2012, the United States and Russia will still retain around 2,000 strategic nuclear weapons each. In addition, a deep level of mistrust remains from the Cold War confrontation and is stoked by continuing disagreements about important security issues, such as NATO expansion, the role of the two powers in Central Asia, and the future of Iraq. These disagreements are unlikely to dissipate quickly, although recent events indicate a willingness on the part of the United States and Russia to explore a new framework for their relationship.

STUDY APPROACH

Our study approach entailed defining a series of phased steps for improving nuclear safety that begin today and go out to roughly 2020. The approach provides both an overall strategy for improving nuclear safety and specific policy steps, on a timeline, to minimize the risk of accidental or unauthorized nuclear use. We also sought to integrate our proposed Nuclear Safety Initiative with two major strategic policies of the United States: improving U.S.-Russian relations and redefining U.S. deterrence needs in light of a rapidly evolving geostrategic environment. At the beginning of the timeline are immediate and near-term steps that could be taken to improve nuclear safety unilaterally or through rapid mutual agreement. These initial steps are designed to build confidence and trust between the two nations. If these steps are successful, more-extensive steps could be taken, in the medium term, to build toward a long-term goal of significant improvements in nuclear safety coupled with a cooperative U.S.-Russian relationship.

One of the difficulties in designing a set of strategies and policies for improving nuclear safety is the broad, cross-cutting nature of the problem. Nuclear safety for Russia and the United States touches on such pivotal and controversial issues as nuclear strategy, the readiness and posture of U.S. and Russian nuclear forces, and the changing nature of U.S.-Russian relations. It also involves multiple federal agencies. Addressing the problem of nuclear safety will therefore require *direct* Presidential leadership and commitment. Within the U.S. government, this could be accomplished two ways. First, it could be done through a National Security Council (NSC) process initiated by a Presidential Directive (PD). The Appendix of this report outlines what might be included in such a PD if President Bush decided to make nuclear safety a priority of his administration. Second, it could be accomplished by the President and a few key advisors making the decisions, thereby avoiding the interagency process altogether—much like the former President Bush did with the unilateral reductions in 1991. Each model has advantages and disadvantages.

What we have done in this report is to present a limited version of the analyses that the Department of Defense and other agencies (or Presidential advisors) might perform to provide the President with various options for reducing the risk of accidental or unauthorized nuclear use. The first item the advisors or agencies would have to consider is the scope of the problem: What is the range of possible scenarios that might lead to accidental or unauthorized nuclear use? Chapter Two covers this first step, exploring the possible scenarios in which an incident of nuclear use might begin. It then goes on to the next step: Identify and assess the underlying factors that might contribute to possible nuclear use (e.g., launch readiness of nuclear forces, perceived vulnerability of nuclear forces or command and control systems to a nuclear first strike, adequacy and reliability of early-warning information, and the amount of time leaders have to decide whether or not to respond to a perceived nuclear attack).

The next step in the process is to define the criteria to be used in evaluating the possible approaches for improving nuclear safety. Because of their uniquely destructive properties, nuclear weapons have both a military and a symbolic role in global affairs, which implies that nuclear safety options (particularly those that change the size, readiness, and operation of nuclear weapons) will affect a broad range of issues. Therefore, an evaluation of the pros and cons of a particular nuclear safety approach should include its effect on U.S.-Russian relations, efforts to prevent the proliferation of weapons of mass destruction, and current U.S. strategies and targeting plans. Chapter Three defines such criteria, as well as criteria directly related to the goal of improving nuclear safety.

Chapter Four examines a wide range of potential approaches, or options, for improving nuclear safety. We selected 10 options for detailed investigation. Each is discussed in a separate section, which includes background on the particular nuclear risks the option is designed to help solve, as well as an introduction to the option itself. Specific technical and operational details on the option are outlined, and the option is evaluated using the criteria established in Chapter Three.

Chapter Five sets forth those 10 options we recommend as the most promising and provides a phased timeline for implementing them. Our recommendations include possible immediate and near-term steps to reduce nuclear dangers and to build confidence and trust in U.S.-Russian relations. Additional steps, in the medium and long term, that move toward the twin long-term goals of a strengthened and cooperative U.S.-Russian relationship and a significantly reduced risk of accidental or unauthorized use, are then described.

All of the potential options are at best only steps toward an ultimate solution of the nuclear safety problem: a U.S.-Russian relationship where neither country views the other as a nuclear threat and postures its nuclear forces accordingly. The current relationship between Britain and France illustrates this end state. Both of these countries are nuclear powers, and they do not have the same interests on all issues. However, neither country would ever consider using nuclear weapons or even military force against the other to settle a dispute. Although this possibility seems remote today for the United States and Russia, the end of the Cold War and the corresponding improvements in relations have created the opportunity for both countries to start the process that might lead to this end state.

Chapter Two

POSSIBLE SCENARIOS FOR ACCIDENTAL OR UNAUTHORIZED NUCLEAR USE

A variety of scenarios could result in the accidental or unauthorized use of nuclear weapons. This chapter begins by analyzing the different types of possible scenarios, giving several examples of each one. Some of the examples are real world experiences that illustrate the types of problems that might lead to accidental or unauthorized use; others are fictional situations that demonstrate how a combination of factors might lead to nuclear use. The chapter then considers all of the underlying factors that might contribute to accidental or unauthorized nuclear use in these scenarios. Only after understanding these factors can solutions be designed to reduce nuclear risks.

POSSIBLE SCENARIOS

Because our goal was to understand the underlying causes for accidental or unauthorized nuclear use, we tried to capture as many types of scenarios as possible. We grouped our results into three basic types of scenarios, as shown in Figure 2.1. We believe this set of eight types encompasses the full range of possible scenarios involving accidental or unauthorized use of nuclear weapons.

Although none of these types of scenarios is very likely, the tremendous destructive capacity of nuclear weapons makes the risk of even one of them occurring unacceptably high. Furthermore, if one were to occur, it is unlikely that it would result directly from a single un-

---

Type I: Unauthorized Launch
- Intentional launch by rogue commander or terrorist

Type II: Launch by Mistake
- Training accident
- System malfunction

Type III: Intentional Launch Based on Incorrect Information
- Malfunction of early-warning system
- Incorrect interpretation of nonthreatening event
- Misperception of nuclear attack by third country or terrorists
- Misperception of accidental nuclear detonation on own territory
- Misinterpretation of simulated training attack as real attack

---

Figure 2.1—Possible Scenarios for Accidental or Unauthorized
Nuclear Use, by Type

derlying factor. It is more likely that a combination of factors would
be involved. We explore the possible scenarios below, grouped according to type.

Type I Scenarios: Unauthorized Launch

Sustained economic hardship and other internal difficulties have
created uncertainty about the stability of Russia's system for maintaining strict control of its nuclear forces and materials. Although the
situation may have improved over the past couple of years, Russia is
a more chaotic place than it was during the Soviet period. During
peacetime, this may not be a serious problem; but under the pressure of a domestic or international crisis, Russia's command and
control of its nuclear forces may face problems.

The two fictional scenarios described next illustrate the kinds of
crises that might lead to an unauthorized launch of nuclear weapons.
The first involves terrorists; the second involves a rogue commander
(i.e., one who takes control of the nuclear forces he commands).

In the first, Russian-Chinese relations become increasingly hostile
between 2002 and 2007. In 2007, increasing nationalism in China

leads to a border dispute with Russia, and both sides mass troops on the border. Russia's conventional forces in the Far East are no match for China's, so Russia heightens the alert level of its nuclear forces. One of the measures it takes is to disperse its mobile ICBM force into the field.

While Russia's mobile ICBMs are in the field, a terrorist organization supported by sympathizers within the Russian military seizes control of a mobile ICBM. The terrorists gain access to launch codes either from sympathizers in the military or by overcoming protection systems during the confusion of a crisis.

The second fictional scenario involves a period of internal political turmoil. Perhaps there is a disputed Russian presidential election, during which military supporters of one candidate seize control of a Russian ballistic missile submarine, ordering all outsiders to stop interfering in Russian domestic affairs. The submarine's captain, fearing that the United States will send attack submarines to destroy his submarine, warns the United States to keep its attack submarines out of his patrol area. He threatens to launch his missiles against the United States if it tries to attack.

Type II Scenarios: Launch by Mistake

In this type of scenario, one or more missiles are launched by mistake during a training exercise or because the command and control system malfunctions. Because the nuclear force training does not involve the use of actual weapons, the probability of an accident is very low. The consequences of such a mistake, however, would be catastrophic.

A real-life example of what can occur during training is provided by an incident that happened on November 9, 1979, when the computer displays at North American Aerospace Command, the Pentagon's National Military Command Center, and the Alternative National Military Command Center all showed a massive Soviet nuclear strike aimed at U.S. nuclear forces and command and control infrastructure. A conference call among the command posts was convened and the launch control centers at the U.S. ICBM facilities were given preliminary warnings that the United States was under nuclear attack. The U.S. continental air defense was also alerted, and at least 10 fighters took off to defend U.S. airspace against possible attacks by Soviet nuclear bombers.[1] Even the President's airborne command center was launched without the President on board.

A crisis was averted when the raw data from the U.S. early-warning launch detection satellites showed no signs that the United States was under nuclear attack. It was later determined that a realistic training tape had been mistakenly inserted into the computer running the nation's early-warning computer programs. To avoid repeating the mistake, the United States trains its personnel in response procedures on a system completely separate from the real missile warning system.

Type III Scenarios: Intentional Launch Based on Incorrect Information

The Type III scenarios (see Figure 2.1) all involve national leaders authorizing nuclear retaliation after being convinced by a faulty reading of the situation that the country is under nuclear attack. This type of scenario could occur either during day-to-day operations or in the midst of a serious crisis. Concern about this kind of scenario has been heightened since the end of the Cold War because of the steady deterioration of Russia's early-warning systems and the increased vulnerability of Russian forces to a disarming first strike by the United States.

Of the various scenarios involving intentional launch based on incorrect information, we illustrate four. The first two are real-life examples of problems associated with malfunctioning early-warning systems. They illustrate the dangerous connection between nuclear forces being kept on high alert, short warning times, and faulty early-warning systems. The second two are fictional situations illustrating what can happen when a nuclear attack by a third party or an accidental nuclear detonation is misperceived as an attack on Russia or the United States.

---

[1] See Geoffrey Forden, *Reducing a Common Danger: Improving Russia's Early Warning System* (Washington, DC: CATO Institute), May 3, 2001.

The first example is an incident that occurred in 1995 when Russia was suddenly faced with an unknown ballistic missile headed toward outer space near Russian territory. Early on the morning of January 25, 1995, Russia's early-warning radar at Olenegorsk detected a ballistic missile streaking toward space from somewhere in the Barents Sea area, where U.S. ballistic missile submarines are known to operate. Within minutes, Russian military commanders had placed Russia's nuclear forces on heightened alert, and for the first time, then President Boris Yeltsin activated his "nuclear football"—the briefcase containing the codes for launching Russia's nuclear forces.[2] He then stood by, waiting to launch Russia's nuclear weapons at the United States if his military commanders determined that the attack was real.[3] After several tense minutes, Russian early-warning satellites determined that no U.S. ICBM had been launched, reinforcing the view of some Russian officials that it was a false alarm. Russian officials reportedly believed that the United States was extremely unlikely to launch a nuclear attack that did not involve using both its Trident submarines and its ICBMs.

The missile turned out to be a NASA sounding rocket, launched from northern Norway toward the North Pole to conduct research on the polar climate. Even though the missile was not heading toward Russia, it could have been a so-called precursor attack in which the United States was exploding a nuclear weapon high in the atmosphere to blind Russia's early-warning radars to a large nuclear attack that would arrive minutes later.[4] This incident occurred in spite of Russia's having been notified more than a month earlier that the launch would take place, as required by agreement. Through a combination of errors and inadequate procedures in Norway, Russia, and the United States, the notification evidently never made it to Russia's

---

[2]See Bruce Blair, Harold Feiveson, and Frank von Hippel, "Taking Nuclear Weapons Off Hair-Trigger Alert," *Scientific American*, November 1997; and Forden, *Reducing a Common Danger*.

[3]Interview with Bruce Blair for *Frontline*, "Russian Roulette," aired February 23, 1999.

[4]Theodore A. Postol, "The Nuclear Danger from Shortfalls in the Capabilities of Russian Early Warning Satellites: A Common Russian-U.S. Interest for Security Cooperation," presentation to Carnegie Endowment for International Peace, February 26, 1999, pp. 17–34.

Strategic Rocket Forces.[5] This episode was what galvanized much of the effort in the United States to reduce the alert rates of nuclear forces and to improve Russia's access to early-warning information.

In the second real-life case, Russia's early-warning system mistakenly indicated that the United States had launched several ICBMs from its continental missile fields.[6] On the evening of September 26, 1983, Lt. Col. Stanislav Petrov, the officer in charge of operating Russia's spaced-based early-warning system, spotted on his computer screen what looked like the launch of several missiles from U.S. ICBM fields. Petrov was getting his data from Russia's new Oko satellites operating in highly elliptical earth orbit. It is unknown whether Petrov realized that the Oko satellites would register false positives when sunlight reflected off high-altitude clouds, a condition that was occurring that day.

Petrov reported that he never passed the warning to his superiors because he quickly concluded it was a false alarm. He said his early-warning system indicated that only five U.S. missiles had been launched, and he thought it highly unlikely that the United States would begin a nuclear conflict with such a limited strike.[7]

The third and fourth situations involve the explosion of a nuclear weapon in or near Russia or the United States, either by terrorists or by accident. The command and control structure of either nation would be severely tested if an actual nuclear explosion occurred on its territory. Quickly, within the first few minutes or hours, national leaders would have to determine whether the explosion was an attack, and, if so, who or what was behind it and how extensive it was. During those first few moments, critical mistakes could be made in trying to arrive at these determinations. Nuclear postures and strategies that require launch on warning are particularly dangerous in these types of scenarios.

---

[5]"Will New Notification Procedures Ensure There Is No Repeat of the 1995 Norwegian Incident?" *De-Alerting Alert*, Issue 1, October 1997.

[6]David Hoffman, "I Had a Funny Feeling in My Gut," *Washington Post*, February 10, 1999, p. A19.

[7]Ibid.

Two fictional examples of such scenarios, in which a nuclear event could lead to an intentional launch based on incorrect information, are given next. These are more likely to occur if a country's early-warning system has large holes in it, its nuclear forces are not survivable, or the incident occurs during a crisis:

1. A terrorist organization explodes a nuclear weapon within a major metropolitan area of one of the countries. The chaos and confusion caused by the nuclear blast leads the national command authority to believe the country is under attack by the other country. It launches a retaliatory nuclear strike.

2. An accidental nuclear detonation occurs in one country while nuclear weapons are being transferred onto a submarine or ICBM. The confusion surrounding the incident leads the national command authority to conclude the country is under attack and to launch a retaliatory nuclear strike against the other country.

FACTORS CONTRIBUTING TO POSSIBLE NUCLEAR USE

If an accidental or unauthorized launch were to occur, it is unlikely that a single factor would have directly led to it. It is more likely that a combination of factors would be needed, including a tense nuclear relationship between the United States and Russia, the presence of a large number of nuclear forces maintained in a manner that makes them ready for immediate launch, and a sudden crisis or accident that creates false impressions and errors in judgment.

We tried to determine all of the underlying factors that might cause any one of the scenarios to occur. Figure 2.2 lists these factors by scenario type. We then pulled out the seven distinct factors in the list (see Figure 2.3). Note that several factors can contribute to more than one scenario.

Using the list of seven factors, one can clearly identify the issue that efforts to improve nuclear safety should address. We used that list to develop potential options for reducing nuclear risk in Chapter Four. Although our focus is on the United States and Russia, the list of factors could be applied to any nuclear state.

Type I Scenarios: Unauthorized Launch
- Inadequate security and control of nuclear forces and weapons
- Nuclear forces kept at high day-to-day launch readiness

Type II Scenarios: Launch by Mistake
- Inadequate security and control of nuclear forces and weapons
- Inadequate training precautions
- Nuclear forces kept at high day-to-day launch readiness

Type III Scenarios: Intentional Launch Based on Incorrect Information
- Nuclear forces kept at high day-to-day launch readiness
- Perceived vulnerability of nuclear forces or command and control systems
- Inadequate early-warning information
- Short decision times
- Deterrence doctrine or posture reliant on launch on warning or launch under attack

Figure 2.2—Contributing Factors for Each Type of Scenario

1. Nuclear forces kept at high day-to-day launch readiness

2. Perceived vulnerability of nuclear forces or command and control systems

3. Inadequate early-warning information

4. Short decision times

5. Deterrence doctrine or posture reliant on launch on warning or launch under attack

6. Inadequate security and control of nuclear forces and weapons

7. Inadequate training precautions

Figure 2.3—Seven Factors Contributing to Possible Nuclear Use

The remainder of this section examines each factor in detail.

Factor 1: Nuclear Forces Kept at High Day-to-Day Launch Readiness

Today, the United States and Russia keep their nuclear forces at high levels of launch readiness. But they have taken very different approaches to how they structure and posture their forces (see description in Chapter One).

This high level of launch readiness in both countries contributes to nuclear dangers in several ways. First, both countries having thousands of nuclear weapons ready to launch within minutes leaves very little time in a crisis for either one to decide whether it is under attack before it must decide how to respond. If the enemy has launched a first strike and the nation's leaders hesitate to make a decision, their forces will be destroyed. This is a particular problem for Russia today, because its forces are smaller and less survivable than the U.S. forces. Russia would have only a small force left if it waited to absorb a U.S. first strike before retaliating. Second, high launch readiness lowers the threshold for accidental nuclear use because it reduces the number of steps needed to launch the missiles. Third, if a rogue commander or a terrorist organization were able to take over ICBM launchers or a ballistic missile submarine, it would have all the components needed for launch except the authorization codes, which are controlled by higher authorities at separate locations. If it could gain access to the launch codes through a well-placed source within the nuclear command authority, it would need to maintain control over the missiles for only a short time in order to launch them.

Factor 2: Perceived Vulnerability of Nuclear Forces or Command and Control Systems

Another factor that contributes to nuclear safety concerns is the awareness of the Russian leadership that its nuclear forces and command infrastructure are vulnerable to a massive U.S. strike utilizing both nuclear and conventional weapons. Such a strike could, in theory, leave Russia with only a very modest retaliatory capability of a hundred or fewer warheads to launch against the United States if it opted to "ride out" a nuclear attack. In response to this concern, Russia has very likely placed much greater reliance on a launch-on-warning strategy that requires it to launch its forces before incoming warheads arrive—i.e., within 25 to 30 minutes if the attack consists of ICBMs coming from the United States, or within as little as 10 to 12 minutes if the attack comes from Trident submarines located a few thousand kilometers off the Russian coast.[8] This strategy could be extremely dangerous during a crisis, because Russia's leaders would have only a few minutes to determine whether what was occurring was a nuclear attack or, instead, a false alarm caused by a malfunction in their early-warning system or a misinterpretation of an actual event. According to some analysts, this is the reason why the January 1995 incident was so dangerous.[9]

There are several interconnected pieces to this problem. First, Russia's economic and financial crisis has severely limited the amount of money going to the armed forces, including Russia's nuclear forces. As a result, Russia has very few forces that would likely survive a surprise attack. Reports indicate that Russia has at most one to two regiments of nine single-warhead road-mobile ICBMs deployed in the field[10] and one to two ballistic missile submarines at sea, each carrying 64 warheads.[11] For a three-month period in 1998, Russia had no ballistic missile submarines at sea, according to one report.[12] Second, the United States continues to posture its nuclear forces in a manner suggestive of an offensive counterforce "damage limitation" strategy—i.e., a strategy to destroy a large portion of Russia's nuclear forces before Russia could retaliate against the United States. Such a strategy requires that a large number of accurate, powerful counterforce weapons be deployed at the ready—meaning they could be launched within minutes—in order to rapidly destroy a high percentage of Russia's nuclear forces. In addition, the United States reportedly continues to patrol its attack submarines near the patrol

---

[8]Blair, Feiveson, and von Hippel, "Taking Nuclear Weapons Off Hair-Trigger Alert"; and Bruce Blair, *Global Zero Alert for Nuclear Forces* (Washington, DC: Brookings), 1995, pp. 43–56.

[9]Blair, Feiveson, and von Hippel, "Taking Nuclear Weapons Off Hair-Trigger Alert"; Blair, *Global Zero Alert*, pp. 43–56; and Postol, "The Nuclear Danger from Shortfalls," pp. 4–6.

[10]Blair, *Global Zero Alert*, pp. 48, 64.

[11]Robert S. Norris and William M. Arkin, "Nuclear Notebook: Russian Nuclear Forces, 2001," *Bulletin of the Atomic Scientists*, Vol. 57, No. 3, May/June 2001, pp. 78–79.

[12]John Downing, "Russian SSBN Patrols Halted for Three Months," *Defense Week*, January 11, 1999, p. 1.

areas of Russia's increasingly strategic submarine force. Third, U.S. conventionally armed air and cruise missile forces in Iraq and Yugoslavia have demonstrated their ability to destroy hardened targets with nonnuclear precision-guided weapons. Russian analysts are concerned that these weapons could be used against Russia's nuclear forces and its command and control infrastructure.

Factor 3: Inadequate Early-Warning Information

The breakup of the Soviet Union and Russia's economic and budgetary problems have caused Russia's early-warning system to badly deteriorate. Analyses conducted by the Congressional Budget Office (CBO) and several U.S. and Russian analysts over the past several years suggest that Russia's early-warning system is in decay, particularly its space-based launch detection system.[13] This trend shows no signs of halting.

Like the United States, Russia relies on a combination of infrared satellite-based sensors and ground-based radars to provide early warning of an attack and to reduce the chances of mistakes. The satellites provide the earliest warning of an attack; they detect the hot exhaust from missiles as they fly into space. The land-based radars track the missiles as they get closer to the target.

Russia's Space-Based Early-Warning System. Russia has relied on two constellations of space-based sensors to detect U.S. missiles as they are launched (see Figure 2.4).[14] The primary one, called Oko, is designed to provide full-time surveillance of rocket launches from ICBM fields in the central United States. For a variety of technological and bureaucratic reasons, Russia has never mastered the technology that would allow its satellites to look straight down at the earth and detect missile plumes (the hot gases that escape from a

---

[13]Geoffrey Forden, "Letter to the Honorable Tom Daschle Regarding Improving Russia's Access to Early-Warning Information"(Washington, DC: CBO), September 3, 1998; Podvig, *Russian Strategic Nuclear Forces* (Cambridge, MA: MIT Press), 2001; and Postol, "The Nuclear Danger from Shortfalls."

[14]See Paul Podvig, "The Operational Status of Russian Space-Based Early Warning System," *Science and Global Security*, Vol. 4, 1994, pp. 363–384, and Podvig, *Russian Strategic Nuclear Forces*, pp. 420–435.

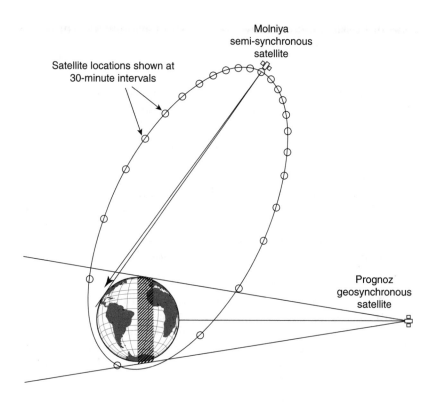

SOURCE: Postol, 1999.

**Figure 2.4—Russia's Molniya (Oko) and Geosynchronous (Prognoz)
Early-Warning Satellites**

rocket motor when it is burning).[15] So Oko satellites must observe
the missiles against the cold background of space, a simpler solution
but one that allows each satellite to view missiles only above a small
portion of the earth. For each satellite to observe U.S. ICBM fields for
as long as possible, Oko satellites are placed in highly elliptical, or so-
called Molniya, orbits.[16] As a result, Russia needs at least six satellites
to provide constant coverage of the central United States, where U.S.

---

[15]Forden, *Reducing a Common Danger*, pp. 11–12; and Postol, "The Nuclear Danger
from Shortfalls," pp. 44–99.

[16]Forden, *Reducing a Common Danger*, pp. 11–12.

ICBMs are based, and those satellites must be replaced every few years.[17]

Furthermore, the Oko constellation provides little, if any, coverage of the oceans, where the bulk of U.S. forces are deployed today on submarines. Compare this with the United States, whose look-down satellites, called the Defense Support Program (DSP), can see a full one-third of the globe because the United States has mastered the technology of looking straight down at the earth.

Since 1984, Russia has complemented the Oko constellation with a system of satellites based in geosynchronous orbits, the same type of orbits the United States uses for its DSP satellites.[18] In theory, satellites in this type of orbit can see the entire one-third of the earth below them. But the placement of the Prognoz satellites suggests their ability to look down is very limited, and that they are instead designed primarily to watch missile fields in the central United States and to provide a redundant system for the Oko constellation.[19] Information from Russian experts indicates that these satellites may be able to look down, but only at areas of the ocean much smaller than would be needed to provide coverage of the broad areas where U.S. ballistic missile submarines could patrol.[20] Today, however, not one of the geosynchronous satellites is operational.[21] As a result, Russia has little, if any, ability to detect the launch of missiles from Trident submarines.

In January 1995, when the incident involving the launching of the sounding rocket from Norway occurred, Russia had nine Oko satellites in orbit at locations suggesting they were all operational, which would have given Russia 24-hour coverage of U.S. ICBM fields.[22] (The top panel of Figure 2.5 shows the nine Oko satellites as they were on January 25, 1995, closely following each other as they made their unique orbit around the earth.) Russia also had one Prognoz

---

[17]Ibid.

[18]Podvig, *Russian Strategic Nuclear Forces*, p. 432.

[19]Forden, *Reducing a Common Danger*, p. 13.

[20]Postol, "The Nuclear Danger from Shortfalls," pp. 44–99.

[21]Podvig, *Russian Strategic Nuclear Forces*, p. 432.

[22]Forden, *Reducing a Common Danger*, p. 12.

January 25, 1995

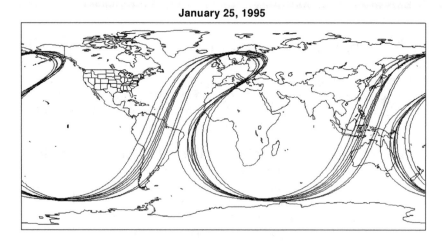

2001

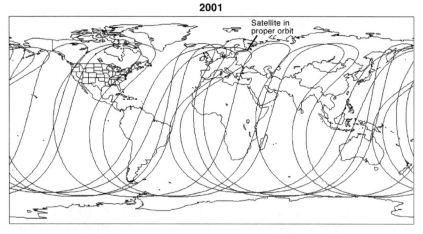

SOURCE: Forden, 2001.

Figure 2.5—Ground Tracks of Russian Oko Satellites on January 25, 1995, and in 2001

satellite in orbit that was in position to monitor U.S. ICBM fields and possibly small areas of the Atlantic Ocean.[23] The around-the-clock

---

[23]Postol, "The Nuclear Danger from Shortfalls," p. 72.

coverage led some analysts to conclude that Russia was able to determine that the sounding rocket was not part of a larger attack because it could see no evidence of launches from U.S. missile fields and its commanders believed that the United States would include its ground-based ICBMs in any attack launched on Russia.[24]

The condition of Russia's space-based early-warning system has deteriorated significantly since 1995.[25] By 1998, only four Oko satellites remained in orbit; and as shown in Figure 2.6 (see the thin line), CBO's analysis found that the constellation could provide at most 17 hours of coverage of U.S. ICBM fields. This assessment is optimistic, however, because it assumes that all the satellites still in orbit continued to function, and Russia is known to allow nonfunctioning

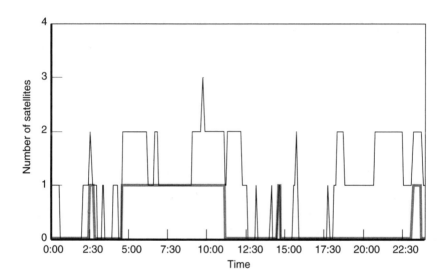

SOURCE: Geoffrey Forden, personal communication.

Figure 2.6—Daily Coverage of U.S. ICBM Fields by Russian Molniya (Oko) Satellites, August 1998

---

[24]Forden, *Reducing a Common Danger*, p. 14.

[25]Ibid, p. 12.

satellites to drift from their original orbits. A look at the orbits of the five Oko satellites in orbit in 2001 (see bottom panel of Figure 2.5) suggests that only one is in the proper orbit and continues to function. What's more, Russian experts indicate that none of Russia's Prognoz geosynchronous satellites were operational in 1998. If this is true, Russia could view U.S. missile fields for only six or seven hours each day in 1998 (see the thick line in Figure 2.6) and had no ability to view Trident patrol areas.

Today, the situation is even worse. Russia has not launched any early-warning satellites since 1998. In addition, a fire in the ground control center for the Oko satellites in May 2001 left the constellation drifting uselessly in space, although control may have been restored recently.[26]

Russia's Ground-Based Early-Warning System. Russia's network of early-warning radars also has significant coverage gaps, although not as severe as those in its space-based early-warning system. Russia relies on an extensive network of ground-based radars as a key element of its early-warning system. But these radars, like those deployed by the United States, are limited in what they can see by the curvature of the earth. They thus cannot detect incoming missiles until much later in their flight than space-based detection systems can, which means they leave much less time for commanders to determine whether there is an attack and to decide whether to retaliate.

Today, there are several problems with Russia's early-warning radar network.[27] First, the second-generation system of large phased-array radars (LPARs) was never completed. Radars planned for construction at Mishalevka and Balkhash were never built, and the one under construction at Krasnoyarsk was torn down because it was in violation of the Antiballistic Missile (ABM) Treaty. In addition, Russia lost one of its completed LPARs following the breakup of the Soviet Union. Opposed to a continued Russian military presence in its now independent state, the Latvian government shut down the Skrunda radar in 1998. All of these incidents have left Russia's current early-

---

[26]See "The Watchers Fall Asleep in Orbit," *Obshchaya Gazeta*, No. 20, May 2001.

[27]This discussion is based on the analysis in Postol, "The Nuclear Danger from Shortfalls," pp. 19–43, 121–138; Podvig, *Russian Strategic Nuclear Forces*, pp. 420–428; and Forden, *Reducing a Common Danger*, pp. 8–9.

warning radar network with two coverage gaps (see Figure 2.7). These gaps are not as severe as those in the space-based system, but they are nonetheless significant. The loss of the Krasnoyarsk radar has left a wide corridor from the Pacific Ocean all the way to Moscow. Trident missiles launched from submarines based in the Pacific could attack through this corridor without being detected by a missile warning radar or a launch detection satellite. The loss of the

SOURCE: Postol, 1999.

Figure 2.7—Gaps in Russia's Early-Warning Radar System

Skrunda radar created a second, narrower corridor that would permit Trident submarines to attack Moscow from the North Atlantic Ocean without being detected.

The Dangers of an Unreliable System. An unreliable early-warning system is dangerous for two fundamental reasons. First, without a clear, accurate picture of what is happening around the globe, Russia may confuse a benign event (such as a space launch) for a nuclear attack, possibly prompting a decision to launch a nuclear strike. Second, without a properly functioning, two-tiered early-warning system, Russia will have less time available to decide about whether to launch a retaliatory response.

Factor 4: Short Decision Times

Nuclear forces at a high state of readiness create the possibility of a nuclear attack with very little warning. This is particularly true for Russia, since the United States could launch counterforce-capable missiles from Trident submarines based a few thousand miles off Russia's coast and strike targets within 10 to 15 minutes of their initial detection by Russian early-warning systems, provided those systems are functioning. Even ICBMs launched from the United States would give Russia only roughly 25 minutes warning before hitting Russian targets, if Russia could detect their launch. These short warning times mean Russian national and military leaders would have an extremely limited amount of time to decide whether the detected event were a real attack and what course of action to take in response. Decision times are even shorter for Russia today—between 0 and 10 minutes—because of its inadequate early-warning system.

During the Cold War, the nuclear postures of the two sides could support this kind of rapid response because of their large and highly survivable forces and their robust early-warning systems. However, even during the Cold War, the very short time between attack detection (when missiles are first detected) and response (the last response time being just before one's warheads are destroyed by the enemy's incoming missiles) required that the entire nuclear retaliation process be carefully orchestrated and all details worked out well in advance. This meant a large percentage of each side's nuclear forces had to be armed and in position to attack on short notice. For the United States, this kind of planning has been made since 1960 in its

Single Integrated Operational Plan (SIOP). The SIOP specifies which nuclear forces and weapons are to be used to attack specific Russian targets in order to fulfill the U.S. deterrence criteria.

Factor 5: Deterrence Doctrine or Posture Reliant on Launch on Warning or Launch Under Attack

When the perceived vulnerability of Russia's nuclear forces and command and control infrastructure is combined with the short warning times provided by Russia's inadequate early-warning system, another route to accidental nuclear use comes into play: the high probability that Russia has adopted a deterrence doctrine reliant upon launch on tactical warning or launch under attack. In other words, the Russians are most likely posturing their nuclear forces in a way that allows for a quick launch, one that occurs before any U.S. nuclear weapons land on Russian territory or just after the first one arrives. Russia's concern is that its forces will be destroyed by weapons launched from Trident submarines close to its coast if it waits for the arrival of U.S. warheads to verify an actual nuclear attack.

If Russia has adopted a launch-on-warning or launch-under-attack strategy, it needs to posture its forces to execute a retaliatory strike within 10 to 15 minutes after detecting an attack, which is just enough time to launch before the first Trident submarine missiles would strike their targets. Russia could also posture its forces to launch within 25 to 30 minutes, which is roughly the time that would elapse between its satellite early-warning system's initial detection of ICBMs launched from the United States and the impact of the ICBM-borne weapons.

According to one report, the Russians have adopted just such a strategy, one that permits them to launch their strategic missiles within 12 minutes of an attack's first detection by Russia's early-warning system.[28] This kind of rapid response allows very little time to verify that early-warning information is correct. It has also been reported that to counter the risk that its command and control system might be destroyed before it could launch its forces, the Russian leadership

---

[28]Blair, Feiveson, and von Hippel, "Taking Nuclear Weapons Off Hair-Trigger Alert."

has developed a system, called the "dead hand," that automatically launches a massive counterattack against the United States if Russia is hit by nuclear weapons.[29] At this point, however, there is serious debate about this system's existence.

Factor 6: Inadequate Security and Control of Nuclear Forces and Materials

Russia's current economic conditions have caused much concern about the security and control of Russian nuclear forces in that they could increase the chances of accidental or unauthorized launch. There are two main dangers. The first is that the security around nuclear facilities and forces will be compromised. For example, a local Russian military commander may become so upset about his troops' poor living conditions that, in protest, he takes control of the nuclear forces he commands—i.e., becomes a rogue commander. Another possibility is that there will be a breakdown of security around nuclear forces, weapons areas, storage areas, or development facilities that will allow terrorists to gain control of nuclear materials or weapons. The poor pay and living conditions of Russian nuclear scientists and soldiers may make them susceptible to bribery and other means of persuasion available to terrorist organizations.

The second main danger is that a breakdown of the command and control process will occur. During the Cold War, Russia and the United States had very effective command and control systems. Russia continues to possess a robust battle management system with numerous backup channels and a variety of redundant means for transmitting launch authorization orders to nuclear forces. However, in a crisis with very short decision times, this system could break down.

Factor 7: Inadequate Training Precautions

During a training exercise, an accidental launch could occur either because the exercise itself causes the launch or because the proper

---

[29]Bruce G. Blair, statement before the House National Security subcommittee, March 13, 1997; and Blair, *Global Zero Alert*, pp. 51–56.

authorities were not properly alerted of the exercise and thus believe they are under attack. A mistake like this occurred on November 9, 1979, when an exercise tape was accidentally inserted into the U.S. early-warning computer system. Since that incident, measures have been taken to minimize the mistakes that might occur during training. For example, the United States now does its nuclear training in a room other than the one housing the early-warning center.

How the United States and Russia train their forces for nuclear conflict is a highly sensitive topic, and only the most-general details are publicly available. Therefore, while we fully acknowledge that this is an important contributing factor to the possibility of an accidental or unauthorized launch, we are unable to offer concrete solutions to this problem.

CONCLUSIONS

We identified eight basic scenarios for possible accidental or unauthorized nuclear use and grouped them into three types: scenarios leading to unauthorized use, scenarios leading to launch by mistake, and scenarios leading to intentional launch based on incorrect information (see Figure 2.1). Our analysis suggests that seven factors could contribute to accidental or unauthorized use in these three categories.

The next chapter sets forth the criteria we used to evaluate the potential nuclear safety options we arrived at based on our seven contributing factors. This sets the stage for the full evaluations and descriptions of the options in Chapter Four.

Chapter Three

CRITERIA FOR EVALUATING NUCLEAR SAFETY OPTIONS

What are the appropriate criteria for judging nuclear safety options? Past studies, both inside the Department of Defense and at the nuclear laboratories, have used a fairly narrow set of criteria, focusing on maintaining the current equilibrium between Russian and U.S. nuclear forces.[1] In their view, a nuclear safety option should be constructed to ensure that neither country gains an advantage over the other once the option is implemented. They also place heavy emphasis on ensuring verification of the proposed measure. These criteria are certainly essential to any evaluation of options for improving nuclear safety, but they are not the only fitting criteria. By focusing narrowly, these studies rejected options even though they may have had merit for reducing the risk of accidental or unauthorized nuclear use.

Our study took a different approach and used a broader set of criteria for evaluating nuclear safety options. We adopted criteria used in previous studies, but we also used criteria that encompass broader issues, such as an option's effect on U.S.-Russian relations, U.S. nonproliferation goals, and counterterrorism goals. In coming to an overall assessment of each option, we also used the rule that no single criterion can lead to an option's disqualification. For instance, a

---

[1]See, for example, Thomas H. Karas, *De-alerting and De-activating Strategic Nuclear Weapons* (Albuquerque, NM: Sandia National Laboratories), SAND2001-0835, April 2001; and Kathleen C. Bailey and Franklin D. Barish, "De-alerting of U.S. Nuclear Forces: A Critical Appraisal," *Comparative Strategy*, Vol. 18, January-March 1999, pp. 1–12.

nuclear safety option that greatly improves U.S.-Russian relations yet is very difficult to verify may still be attractive—as long as it does not significantly undermine the U.S or Russian deterrent. Nor must one option do everything at once. Partial steps may have a significant effect on improving the tone of U.S.-Russian relations, thereby reducing the chances of accidental nuclear use and creating the foundation for nuclear safety initiatives.

In our view, the primary goal should always be kept in mind: Strengthen nuclear safety by reducing the risk of accidental or unauthorized launch of nuclear weapons and improving U.S.-Russian relations while simultaneously preserving mutual deterrence at a level appropriate to current needs. In line with this goal, we used the following eight criteria to evaluate nuclear safety options:

1. *Contribution to reducing the risk of nuclear use.* How does the option specifically reduce the risk of nuclear use? What contributing factor(s) does the option address? Does it have a major or minor positive (or negative) effect on the potential for future nuclear use? For example, does it increase the amount of time that national leaders would have to decide if a detected event is an attack? Does it improve the survivability of nuclear forces? This criterion also tries to qualitatively capture how much a particular option will improve nuclear safety. For example, how much additional time does the option provide? What additional pieces of information might be provided to national decisionmakers during these critical minutes or hours?

2. *Effect on current U.S. strategies and targeting plans.* What is the option's effect on current U.S. strategies and targeting plans? Would nuclear strategies and targeting plans need to be altered or modified if the option were implemented?

3. *Effect on U.S.-Russian political relations.* Does the option enhance the level of confidence and trust in the U.S.-Russian relationship? Does it diminish the importance of nuclear weapons in the relationship?

4. *Effect on other major international actors (China, Europe, etc.).* How does the option affect other major international actors? Does it increase, have no effect on, or decrease the probability of nuclear use—deliberate or accidental—by these other actors? Will

participation by other actors enhance or detract from the option's effect on the potential for future nuclear use?

5. *Effect on prospects for achieving nonproliferation and counterterrorism goals.* Does the option enhance or detract from efforts by the United States to achieve its nonproliferation and counterterrorism policy goals? Will nuclear weapons in Russia and the United States be more secure or less secure if the option is implemented?

6. *Feasibility and affordability.* Is implementation of the option technically feasible? What effect will its implementation have on military operations? How expensive will implementation be? Are any of these factors great enough to discourage U.S. or Russian participation?

7. *Effect on incentive to strike first with nuclear weapons.* Does the option strengthen or weaken first-strike stability? After the option is implemented, will nuclear forces be more survivable or less survivable than they are today? In a crisis, does one country or the other gain a significant advantage if there is a race to rearm?

8. *Ability to monitor or verify implementation of the option and effect of cheating.* How easily can the option's implementation be monitored or verified? If cheating by either country occurs, how will it affect nuclear stability? Will cheating let the offending nation gain the ability to launch a disarming first strike?

In the next chapter, we use these eight criteria to evaluate a range of options for improving nuclear safety.

Chapter Four

OPTIONS FOR IMPROVING NUCLEAR SAFETY

Many different approaches could be used to address the factors that could contribute to nuclear use. In this chapter, we first discuss a wide range of possible options for addressing the factors we identified in Chapter Two as contributing to the accidental or unauthorized use of nuclear weapons. Some of these approaches could be implemented unilaterally, some cooperatively, and some either way. We then focus on ten specific options for addressing the nuclear risk problem, develop them fully, and evaluate them carefully. We emphasize the details of each option as much as possible in our evaluation, because the way in which an option is implemented is likely to affect its success. We discuss our recommendations about which options to pursue and in which order in the next chapter.

FINDING SOLUTIONS TO THE POTENTIAL CAUSES OF NUCLEAR USE

The analysis in Chapter Two suggests that accidental or unauthorized use of nuclear weapons could be caused by a number of underlying factors—from nuclear forces being kept at high levels of alert, to inadequate early-warning information, to inadequate security and control of nuclear forces. But what can be done to eliminate or control these contributing factors and thereby reduce the chances of nuclear use? Or if these factors cannot be addressed directly, can something be done to mitigate the results of nuclear use to some degree?

Table 4.1 shows the wide range of possible approaches we explored as ways to address each contributing factor. For example, in the case

Table 4.1

Illustrative Approaches for Addressing Factors That Could Contribute to Accidental or Unauthorized Nuclear Use

Contributing Factors	Illustrative Approaches[a]
Nuclear forces kept at high day-to-day launch readiness	• Pull back SSBNs from forward areas • Keep attack submarines out of bastions • Deploy missile defenses • Adopt a new deterrence strategy • Install DAL mechanisms • Reduce launch readiness (for part or all of force)
Perceived vulnerability of nuclear forces or command and control systems	• Reduce forces • Pull back SSBNs from forward areas • Keep attack submarines out of bastions • Remove W-88 warheads • Adopt a new deterrence strategy • Reduce counterforce potential • Transfer command and control technology • Improve survivability of forces and command and control systems
Inadequate early-warning information	• Provide shared access to reliable early-warning information • Provide funding to fill in Russian space and radar network • Deploy sensors on silos
Short decision times	• Pull back SSBNs from forward areas • Keep attack submarines out of bastions • Reduce launch readiness (for part or all of force) • Adopt a new deterrence strategy • Install DAL mechanisms • Provide shared access to reliable early-warning information • Provide funding to fill in Russian space and radar network • Improve survivability of forces and command and control systems
Deterrence doctrine or posture reliant on launch on warning or launch under attack	• Reduce counterforce potential • Pull back SSBNs from forward areas • Keep attack submarines out of bastions • Transfer command and control technology • Adopt new deterrence strategy • Reduce launch readiness (for part or all of force) • Provide shared access to reliable early-warning information • Deploy sensors on silos
Inadequate security and control of nuclear forces	• Transfer command and control technology • Deploy missile defenses • Install DAL mechanisms • Improve morale and personnel reliability programs
Inadequate training precautions	• Conduct training exercises on isolated systems • Install DAL mechanisms • Deploy missile defenses

NOTE: SSBN = ballistic missile submarine; DAL = destruct after launch.
[a]These approaches are not in any particular order.

of high levels of day-to-day launch readiness, we examined approaches that would reduce the vulnerability of nuclear forces, thereby eliminating an important incentive for keeping forces ready to launch at a moment's notice. Two options in this case are to keep attack submarines out of ballistic missile submarine bastions and to keep ballistic missile submarines far away from their targets. We also studied ways to reduce the effects of nuclear weapon use, including deploying missile defenses and installing mechanisms on missiles that would destroy them or their warheads if the launch were in error or without authorization. In addition, we examined options aimed at changing deterrence strategies in order to reduce a country's reliance on damage limitation, which, in turn, could reduce the need to keep large numbers of prompt counterforce weapons ready to launch quickly. Finally, we examined options for reducing the launch readiness of nuclear forces directly.

Another contributing factor—the perceived vulnerability of nuclear forces and command and control systems—could be addressed by reducing the nuclear forces that threaten them or by altering the way that the most-threatening forces are postured. Examples include reducing inequities in the numbers of deployed nuclear forces, removing the counterforce potential of nuclear forces, and keeping attack submarines and ballistic missile submarines away, as discussed above. Adopting a new deterrence strategy that does not require large, rapid reaction forces can also help, particularly if it allows changes in the way that nuclear forces are postured and operated that are visible and perceived as less threatening. Of course, perceived vulnerability can also be improved by deploying and operating forces in such a way that they are more survivable on a day-to-day basis. More-survivable forces will not help, however, if command and control systems remain vulnerable. So improvements in the survivability of these systems, including technology transfer, might be considered to improve nuclear safety.

Inadequate access to reliable and accurate early-warning information can be addressed by sharing early-warning information, taking concrete steps to fill holes in Russia's radar and satellite early-warning systems, or developing complementary early-warning systems, such as deploying sensors outside ICBM silos that detect launches.

Short decision times can be addressed by improving access to early-warning information, increasing the survivability of nuclear forces, adopting a deterrence strategy that does not require prompt responses, or reducing the launch readiness of nuclear forces.

Another contributing factor—deterrence doctrines reliant on launch on warning or launch under attack—can be addressed by many of the approaches outlined above for improving the survivability of nuclear forces and command and control systems, particularly if having more-survivable forces encourages a country to back away from prompt launches. Improving access to early-warning information can also be helpful if it is done in ways that extend warning times.

Inadequate security and control of nuclear forces can be addressed by transferring command and control technology, improving the morale of personnel in sensitive positions, and improving personnel reliability programs. If controls fail and a launch occurs, destruct-after-launch (DAL) systems and missile defenses may be able to mitigate the effects of the launch.

The final factor—inadequate training precautions—can be addressed through programs designed to correct weaknesses in existing training programs, including conducting training exercises only on isolated systems that have been disabled or do not carry live nuclear warheads. DAL and missile defense systems could be used as a last resort to address the effects of an accidental launch.

As the preceding discussion indicates, many of the approaches we examined can address more than one contributing factor. In fact, there are only 14 different approaches in Table 4.1. These are quite general, however: "Improve access to early-warning information," "reduce launch readiness," and "deploy missile defenses" can, for example, each be implemented in many ways. Moreover, the specifics of how an approach would be implemented and monitored are central to its success in reducing nuclear risk. A poorly designed approach might even increase the risk of accidental or unauthorized use.

To examine these issues thoroughly, we created a set of ten specific nuclear safety options that represent concrete ways to design a number of the approaches listed in Table 4.1:

1. Provide assistance for improving Russia's early-warning radars or satellites.

2. Establish a joint, redundant early-warning system by placing sensors outside U.S. and Russian missile silos.

3. Immediately stand down all nuclear forces to be eliminated under the 2002 Moscow Treaty.

4. Pull U.S. strategic ballistic missile submarines away from Russia.

5. Keep U.S. attack submarines away from Russia.

6. Remove W-88 warheads from Trident missiles.

7. Reduce day-to-day launch readiness of 150 ICBMs in silos.

8. Reduce day-to-day launch readiness of all nuclear forces.

9. Install destruct-after-launch (DAL) mechanisms on ballistic missiles.

10. Deploy limited U.S. missile defenses.

These ten options were chosen for three reasons: (1) In an initial assessment, some options appeared to have a great deal of promise for reducing nuclear dangers and improving nuclear safety. In our judgment, they thus deserved further exploration. (2) Some options had been proposed by one or more analysts investigating the problem of accidental and unauthorized launch. We wanted to evaluate these approaches against our decision criteria for evaluating nuclear safety options (see Chapter Three). (3) Some options had been rejected by other studies as difficult to verify or as seeming to place Russia or the United States at a disadvantage due to the historical asymmetries in the countries' nuclear forces.[1] In our view, these

---

[1]See, for example, Kathleen C. Bailey and Franklin D. Barish, "De-alerting of U.S. Nuclear Forces: A Critical Appraisal," *Comparative Strategy*, Vol. 18, January-March 1999; National Institute for Public Policy, *De-alerting Proposals for Stragetic Nuclear Forces: A Critical Analysis* (Fairfax, VA: National Institute for Public Policy), June 23, 1999; Michael W. Edenburn et al., *De-alerting Stragetic Ballistic Missiles* (Albuquerque, NM: Sandia National Laboratories), Cooperative Monitoring Center Occasional Paper/9, SAND 98-0505/9, March 1999; and Thomas H. Karas, *De-alerting and De-activating Strategic Nuclear Weapons* (Albuquerque, NM: Sandia National Laboratories), SAND2001-0835, April 2001.

studies seemed overly concerned with maintaining the Cold War equilibrium of Russian and U.S. nuclear forces and thus outdated considering the vast changes in U.S-Russian relations since 1991. We wanted to reconsider these options, taking the broader view offered by our criteria so as to take into account their effects on larger issues, such as U.S.-Russian relations and U.S. nonproliferation and counterterrorism goals, as well as their overall feasibility and cost.

There are three important types of options that we did not explore in detail but whose importance we do not want readers to overlook: adopting new deterrence strategies, transferring command and control technology, and improving training practices. Specific information about nuclear deterrence strategies and doctrines is a closely held secret. Similarly, information about command and control systems, their status, and the technology involved is kept secret by both countries, as are the details of training procedures. All of this makes it difficult to develop detailed options. Nevertheless, new deterrence strategies are important to achieving significant progress in nuclear safety, and their significance is discussed in general terms for several of the options. It will also be critical for both Russia and the United States to maintain very strong and survivable command and control systems and to take whatever steps are necessary, including sharing technology, to ensure these systems' viability. And both nations must remain vigilant in their efforts to avoid training accidents.

The remainder of this chapter is devoted to describing and evaluating our 10 specific nuclear safety options. Note that many of the options are presented as two separate approaches that could be adopted, the choice of one over the other usually relating to the degree of allowable or available monitoring.

OPTION 1: PROVIDE ASSISTANCE FOR IMPROVING RUSSIA'S EARLY-WARNING RADARS OR SATELLITES

This option is a possible solution for two contributing factors:

- Inadequate early-warning information
- Short decision times

Table 4.2 summarizes our evaluation of Option 1 in terms of the nuclear safety criteria. For this option, two separate approaches are offered, as discussed below.

Background

A high-quality, reliable early-warning system can reduce the risk of unintended nuclear use in two fundamental ways. It can improve the quality of early-warning information, thereby ensuring that both countries have a clear, accurate picture of what is happening around the globe, and it can increase the warning time available to each country before it must make a decision to launch a counterattack.

Chapter Two discusses the worrisome state of Russia's satellite– and radar-based early-warning systems. There are several ways the

Table 4.2

Evaluation of Option 1

Nuclear Safety Criterion	Option: Provide assistance for improving Russia's early-warning radars or satellites	
	Approach 1: Provide funding for Russia's radar network	Approach 2: Provide funding and technology for Russia's satellite network
Contribution to reducing the risk of nuclear use	Very positive	Very positive
Effect on current U.S. strategies and targeting plans	N/A	N/A
Effect on U.S.- Russian political relations	Very positive	Very positive
Effect on other major international actors	N/A	N/A
Effect on prospects for achieving nonproliferation and counterterrorism goals	N/A	N/A
Feasibility and affordability	Negative	Negative
Effect on incentive to strike first with nuclear weapons	Positive	Positive
Ability to monitor or verify implementation of the option and effect of cheating	N/A	N/A

United States can provide funding and technology to help Russia overcome the shortcomings in its early-warning network. First, to varying degrees the United States can help fund Russian efforts to launch new satellites or to construct early-warning radars. This could range from paying Russian scientists to continue researching early-warning satellites, to fully funding the development, construction, and launch of Russian early-warning satellite systems. Second, there are various levels of technology the United States can provide. The United States is already involved in a joint venture with Russia—known as the Russian-American Observational Satellite (RAMOS) program—that involves the efforts of American and Russian scientists to improve early-warning technology for both countries.[2]

Specifics

Approach 1: Provide Funding for Russia's Radar Network. Approach 1 illustrates one potential avenue for improving Russia's early-warning system: the United States provides funding to help Russia build two new Pechora-class early-warning radars. The first would be deployed in Belarus, a close ally of Russia, to plug the hole caused by the loss of the Skrunda radar. The second would be deployed in Russia's far east to fill the gap that would have been filled by the now disassembled Krasnoyarsk radar. In each case, the new radar would be the same as Russia's existing phased-array radars, which are the newest type of early-warning radar Russia possesses.

The two new radars would increase the warning time for Russian leaders for a U.S. attack coming through these corridors by 5 to 15 minutes (the time range reflects the variety of possible launch locations and aim points of the attacking Trident missiles). If Russia were to deploy a global space-based system like the DSP, the radars would not necessarily add any warning time, but they would improve the quality of the early-warning information by giving more-precise

---

[2]For an extensive discussion of other ways to improve Russia's early-warning system, see Geoffrey Forden, "Letter to the Honorable Tom Daschle Regarding Improving Russia's Access to Early-Warning Information" (Washington, DC: CBO), September 3, 1998; and Geoffrey Forden, "Letter to the Honorable Tom Daschle on Further Options to Improve Russia's Access to Early-Warning Information" (Washington, DC: CBO), August 24, 1999.

location and tracking information and adding the ability to confirm whether a possible attack is real or not.

Approach 2: Provide Funding and Technology for Russia's Satellite Network. Another way to improve Russia's access to reliable, accurate early-warning information is to provide funding and technical assistance for Russia's early-warning satellite system. According to Russian scientists, six completed Oko early-warning satellites are sitting on the ground waiting to be launched into space.[3] The United States could pay to have these highly elliptical earth-orbiting Oko satellites launched on Russian launch vehicles. If these satellites were deployed, Russia would once again have 24-hour coverage of U.S. ICBM fields (but would not have coverage of the areas where Trident submarines operate). This option would cost approximately $160 million, according to the Congressional Budget Office (CBO).[4] However, it would be only a partial solution to Russia's early-warning problems, because Oko satellites are effective for only three years due to their highly elliptical orbits.[5] Thus, additional launches would be required to keep Russia's early-warning system operational over the next 10 years.

Another possibility is for the United States and Russia to accelerate their current joint research efforts to improve the effectiveness of early-warning technology—i.e., the RAMOS program. One area of research RAMOS is investigating is how to improve a satellite's ability to filter out reflected sunlight from clouds, snow, ice, and oceans—a major cause of clutter for space-based early-warning systems. If this research is successful, Russia will be able to deploy geosynchronous satellites with capabilities much greater than those of either its current Oko satellites or its now dysfunctional Prognoz geosynchronous satellites. Russia would then be able to achieve global coverage with only three satellites.[6]

---

[3]Geoffrey Forden, *Reducing a Common Danger: Improving Russia's Early Warning System* (Washington, DC: CATO Institute), May 3, 2001, p. 15.

[4]Forden, "Letter to the Honorable Tom Daschle, 1998," pp. 1–14.

[5]Forden, *Reducing a Common Danger*, p. 12.

[6]Forden, *Reducing a Common Danger*, p. 16.

The United States could accelerate Russia's development of next-generation geosynchronous satellites by letting Russia import sensors and other components for satellite construction from the West. This transfer of technology would allow Russia to speed up its development process by quickly making available key satellite components that otherwise will take years to develop and manufacture. To transfer the technology, the United States would have to waive certain export restrictions on sensitive space-based technologies—restrictions put in place to prevent adversaries from gaining information about U.S. early-warning system capabilities and access to sensitive technologies.

Evaluation

Contribution to Reducing the Risk of Nuclear Use. Improving the ground- or space-based portions of Russia's early-warning system will greatly diminish the potential for an accidental launch based on incorrect or inadequate information. Enhancement of the early-warning system should provide Russian leaders with an additional 10 to 25 minutes to determine whether they are under nuclear attack by the United States. (See Table 4.3, which compares the effect of each option on decision time, time to re-alert, and time to detect cheating.) It also would provide better-quality and more-reliable early-warning information, which should give national leaders more confidence in their deterrent capability, thereby adding several more minutes to the time they have to determine what kind of response, if any, is appropriate. A global space-based early-warning system would also provide confirmation that Russia was not being attacked by countries other than the United States.

Effect on Current U.S. Strategies and Targeting Plans. Improvements to Russia's early-warning system will have only a minimal effect on current U.S. strategies and targeting plans, but they will reduce the chances of Russian accidental use, which is an important part of stability. It is possible, however, that the transfer of sensitive space-based technology and U.S. help in constructing ground-based radars will improve Russia's nuclear war-fighting capabilities.

First, a technology transfer might allow Russia to understand the limitations and weaknesses of the U.S. early-warning system. Russia also might transfer what it learns to third countries, such as China or

North Korea. This issue has become more important because of Russia's technical assistance to Iran's space program. However, the United States already shares early-warning information with key allies and has a system in place to protect sensitive information. Furthermore, the technology the United States would transfer is from the 1970s, and many of the key materials and systems will be changed in the new Space Based Infrared System High (SBIRS-High) that the United States plans to deploy within the next decade.

Second, if the United States helps Russia improve its ground-based radars, Russia may be able to devise a primitive national anti-ballistic missile (ABM) system. In fact, during the 1970s, the United States accused Russia of violating the ABM Treaty by building the Pechora-type ground radars in the first place. Although it is technically possible that Russia could develop a very limited missile defense, the cost involved seems prohibitive given Russia's economic condition, and the quality of those long-wavelength systems limits their use for missile defense. Finally, additional ground-based radars and improved data processor technology might improve Russia's air-to-air or air-to-ground missile capabilities. However, Russia already excels in this area, and any technology the United States transfers is unlikely to be an improvement.

Effect on U.S.-Russian Political Relations. Joint U.S.-Russian efforts to upgrade early-warning systems are likely to improve relations between the two nations. Greater levels of technology transfer would indicate a new level of U.S. trust and confidence in Russia, placing Russia among a select group of countries (such as Britain and France) with which the United States is willing to share sensitive information.

This option will have only a minimal effect on diminishing the importance of nuclear weapons in the U.S.-Russian relationship. In fact, in a perverse sense, it emphasizes the importance of nuclear weapons by acknowledging that an attack is possible.

Effect on Other Major International Actors (China, Europe, etc.). Option 1 will have a minimal effect on other major international actors. Its only effect is likely to be an indication that Russia and the United States are working together to limit the potential for an accidental nuclear conflict.

Table 4.3

Effects of the Options on Time to Re-alert, Time to Detect Cheating, and Decision Time

Option	Time to Re-alert		Time to Detect Cheating		Extension in Deci- sion Time
	First Missile	Entire Arsenal	First Missile	Entire Arsenal	
1. Help improve Russia's early-warning system by:					
Approach 1: Provide funding for Russia's radar network	N/A	N/A	N/A	N/A	15–25 minutes, 24 hours a day[a]
Approach 2: Provide funding and technology for Russia's satellite network	N/A	N/A	N/A	N/A	5–10 minutes
2. Establish early-warning system by deploying sensors on silos	N/A	N/A	Minutes	Few minutes	15–20 minutes, 24 hours a day
3. Immediately stand down all nuclear forces to be eliminated under Moscow Treaty	Day	Months	Month	Months	(b)
4. Pull Tridents away from Russia:					
Approach 1: No monitoring	Week	Month	—	Many months	10–15 minutes
Approach 2: Monitoring	Week	Month	5–10 weeks	2–3 weeks	10–15 minutes
5. Keep U.S. attack submarines away from Russia	Week	Month	Week	Months	Hours to days
6. Remove W-88 warheads from Trident missiles	Day	Weeks	Months	Months	10–15 minutes
7. Reduce launch readiness of 150 silo-based missiles:					
Approach 1: Enhanced START monitoring	Day	Weeks	Months	Months	(b)
Approach 2: Continuous monitoring	Day	Weeks	Minutes	Minutes	(b)

Table 4.3 (continued)

Option	Time to Re-alert First Missile	Time to Re-alert Entire Arsenal	Time to Detect Cheating First Missile	Time to Detect Cheating Entire Arsenal	Extension in Decision Time
8. Reduce launch readiness of all forces:					
Approach 1: Enhanced START monitoring	Day	Weeks	Months	Months	Days to weeks
Approach 2: Continuous monitoring	Day	Weeks	Minutes	Minutes	Days to weeks
9. Install DAL mechanisms	N/A	N/A	N/A	N/A	(c)
10. Deploy limited missile defenses of the U.S.	N/A	N/A	N/A	N/A	N/A

NOTES: N/A = not applicable; SSBN = ballistic missile submarine; DAL = destruct after launch.

[a]This option ensures coverage of all attack corridors to Russia. It closes gaps that U.S. submarine-launched missiles could exploit. It does not, however, extend the warning time against attacks by U.S. ICBMs. Nor does it extend warning time if Russia deploys a space-based system that provides global coverage.

[b]This option does not affect the launch readiness of a significant portion of the force and therefore has no effect on decision time.

[c]Although a DAL system would provide some time to abort an attack by ballistic missiles (10–20 minutes in the case of a midcourse system), it would not extend decision time in the traditional sense because no leader would launch first and plan to recall later if he were wrong—the risks would be too great.

Some defense experts have suggested that a global early-warning system be set up to provide information on ballistic missile launches around the world. This could be particularly useful in South Asia, because India, Pakistan, and China have no early-warning systems, and the possibility of misinterpreted information leading to nuclear conflict in the region is real. It might be possible for the United States and Russia to jointly provide information to an early-warning center in each country. The problem with this idea is that the countries may not believe the information, particularly during a crisis with the potential for nuclear use. Still, this is one option that would allow the United States to expand its nuclear safety efforts beyond Russia, something likely to become increasingly important as more nations acquire nuclear weapons.

Effect on Prospects for Achieving Nonproliferation and Counterterrorism Goals. Joint U.S.-Russian efforts to improve early warning will probably have little direct effect on U.S. nonproliferation and counterterrorism goals. To the extent that the option provides funding and employment for Russian scientists, however, the United States is making it less likely that those scientists will succumb to financial pressure to work for rogue nations looking for assistance developing weapons of mass destruction.

Feasibility and Affordability. Both approaches for this option are technically feasible, since Russian scientists over the past 30 years have shown a great deal of scientific and technical ability with regard to early-warning systems. What Russia is currently lacking is the funding to implement technological fixes.

Some analysts believe Russia's fiscal problems are exaggerated. They contend that Russian leaders would find the means to fund an early-warning system if they believed it to be vital to their security. Other analysts believe Russia has put considerable effort into keeping its early-warning system operational. For example, in early 1998, Russia's Missile and Space Forces opened a Far East satellite control center that can be used to control geostationary satellites over the Pacific Ocean. This would seem to be an indication that Russia is preparing to launch another generation of geostationary early-warning satellites.[7]

If fully implemented, both approaches would cost a considerable amount of money. For example, if the United States chose to implement Approach 2 by completely funding all of Russia's early-warning satellite programs, including launching the current Oko satellites, the total cost would be around $1.3 billion over the next five years.[8]

Effect on the Incentive to Strike First with Nuclear Weapons. Improved early-warning systems should increase first-strike and crisis stability. If Russia's leaders receive timely and accurate information during a crisis, they are less likely to launch nuclear weapons in error. Early-warning systems will not affect the number of nuclear

---

[7]See, for example, Pavel Podvig (ed.), *Russian Strategic Nuclear Forces* (Cambridge, MA: MIT Press), 2001, pp. 577–578.

[8]CBO estimate in Forden, "Letter to the Honorable Tom Daschle," 1998.

forces available for retaliation after a first strike, but they will provide additional time and information and therefore should increase national leaders' confidence in their deterrent capabilities.

Ability to Monitor or Verify Implementation of the Option and Effect of Cheating. Since the purpose of Option 1 is to provide Russia with improved early-warning information, the system will be completely under Russian control. Therefore, there is no need for verification or monitoring, and it is hard to imagine how cheating could occur in either country.

OPTION 2: ESTABLISH A JOINT, REDUNDANT EARLY-WARNING SYSTEM BY PLACING SENSORS OUTSIDE U.S. AND RUSSIAN MISSILE SILOS

Option 2 is a possible solution for three contributing factors:

- Inadequate early-warning information

- Short decision times

- Deterrence doctrine or posture reliant on launch on warning or launch under attack

Table 4.4 summarizes the evaluation of this option.

Table 4.4

Evaluation of Option 2

Nuclear Safety Criterion	Option: Establish an early-warning system by placing sensors on silos
Contribution to reducing the risk of nuclear use	Very positive
Effect on current U.S. strategies and targeting plans	N/A
Effect on U.S.- Russian political relations	Positive
Effect on other major international actors	N/A
Effect on prospects for achieving nonproliferation and counterterrorism goals	N/A
Feasibility and affordability	Positive
Effect on incentive to strike first with nuclear weapons	Positive
Ability to monitor or verify implementation of the option and effect of cheating	Positive

Background

Option 2 takes an unconventional approach to improving access to reliable early-warning information. It establishes an early-warning network by placing sensors outside each silo so that the United States and Russia would each know immediately if the other launched its intercontinental ballistic missiles (ICBMs). The sensors would be sampled frequently to make sure the missiles were still in their silos. This system would provide nearly instant warning of an attack (or, more important, confirmation that there were no attack). In fact, the warning would arrive sooner than that from space-based sensors, which must wait a minute or so for the missiles to break through clouds. However, this approach requires a moderate degree of cooperation between the two countries.

The idea of such an instantaneous, cooperative early-warning system is not new. The concept was proposed in the 1980s by Richard Garwin,[9] and it was revived more recently by several prominent academicians at the Kurchatov Institute in Moscow, Russia's premier civilian nuclear science establishment.[10] These Russians propose that a cooperative monitoring system originally conceived as a joint approach to monitoring stored nuclear materials taken from dismantled weapons be adapted for early-warning purposes. This monitoring system has been developed over the past seven years for materials storage by a collaboration of the Kurchatov Institute, Arzamas-16 (one of Russia's nuclear weapons laboratories), and Sandia National Laboratories (part of the U.S. Department of Energy's nuclear weapons laboratories). Most of the funding has come from the United States as part of the Materials Protection, Control and Accounting program funded by the U.S. Department of Energy. The monitoring system's goal is to assure the safety, security, transparency, and international accountability of the materials.

---

[9]See, for example, Richard L. Garwin, "Post-START: What Do We Want? What Can We Achieve?" testimony to the U.S. Senate Committee on Foreign Relations, February 27, 1992; and International Foundation for the Survival and Development of Humanity, *Reducing the Dangers of Accidental and Unauthorized Nuclear Launch and Terrorist Attack: Alternatives to a Ballistic Missile Defense System* (San Francisco, CA: International Foundation for the Survival and Development of Humanity), January 1990.

[10]Evgeny Velikhov, Nikolai Ponomarev-Stepnoi, and Vladimir Sukhoruchkin, "Mutual Remote Monitoring," briefing, Kurchatov Institute, Moscow, January 26, 1999.

The nuclear materials monitoring system consists of a series of modules that collect data from a variety of sensors—door switches, motion detectors, fiber optic seals, and video cameras—and transmit the data to a nearby system that collects data from all modules at the facility. These data then go to two places: a monitoring center in the host country (the one being monitored) and, by satellite link or telephone line, to the country doing the monitoring. The data can be transmitted at regular intervals or by request. Sensors can also send information about their state of health. Sandia has expended significant effort to develop systems that are difficult to tamper with or fool, although more work remains to be done in this area. The system uses unique authentication codes to ensure that the sensors are not tampered with. In addition, the system itself is designed to resist tampering. To make sure that data are not tampered with, they are encrypted. The sensors attached to each data module are designed to complement each other to guard against tampering. For example, if infrared motion detectors sense a person in the room, they can trigger the video camera and send an alarm to the monitoring station. Components of this system, including the data collection and transmission modules and some newly designed sensors, have been developed and tested in a series of experiments conducted by Sandia and the two Russian laboratories. A unique feature of this program is that each country develops its own technology for application to its own storage facilities.

The Kurchatov proposal for a cooperative early-warning system is quite broad; it would adapt this system to monitor the status of all strategic nuclear forces. The advantage of this approach is that it is based on systems that have already been developed cooperatively and subjected to some field testing. A module connected to a variety of sensors would be deployed on each silo, mobile missile, submarine missile tube, and, for bomber weapons, on each magazine. It would detect that a missile is being launched or a weapon is being loaded onto a bomber. The module would collect data from the sensors and transmit those data to the central collection system every minute or so via radio. The central facility would send the data to the host and monitoring countries by either telephone line or the Internet.

The resulting system would provide current information about the status of each missile and magazine. If a missile is launched or a silo door opened, the observing party will know about it within one minute. What is even more important in the context of avoiding accidental launches, however, is that the data will indicate that an attack has *not* been launched. If used to complement the traditional space- and ground-based early-warning systems that both countries operate today, the launcher-based system proposed by scientists at the Kurchatov Institute will provide important confirmations of events observed by satellite or by radar. The system will also give nearly instantaneous warning of an attack, thereby increasing the time available for decisionmakers to determine an appropriate response. A particular appeal of this cooperative system is that it can be used as the primary early-warning system, possibly providing a low-cost alternative to improving Russia's existing early-warning systems.

Specifics

Option 2 applies the proposed Kurchatov system to silo-based ICBMs because they are the easiest forces to monitor. Placing monitoring equipment on mobile missiles and submarines can jeopardize the stealth that such platforms rely on for survivability. The main reason to use silo-based sensors instead of improving Russia's space-based early-warning system is that Russia believes the United States will use its ICBMs in a first strike—a reasonable assumption given that U.S. silo-based missiles are widely viewed as powerful and accurate but vulnerable to a retaliatory strike by Russia. The United States holds the same view about the role of ICBMs in Russia's nuclear war plans. This emphasis on U.S. ICBMs for signaling a first strike might lose some of its effectiveness in the future, however, because a larger portion of U.S. warheads will be based at sea.

The basic approach is to detect a launch by monitoring the physical signs that accompany a missile launch (see Figure 4.1). This can be accomplished by using seismic sensors that detect the strong vibrations caused by a missile leaving the silo and heat sensors that detect the hot exhaust gas from the rocket motor. The sensor suite and data collection module could protect themselves against tampering with an infrared motion detector and a video camera. The modules and sensors would be installed by inspectors during so-called baseline inspections.

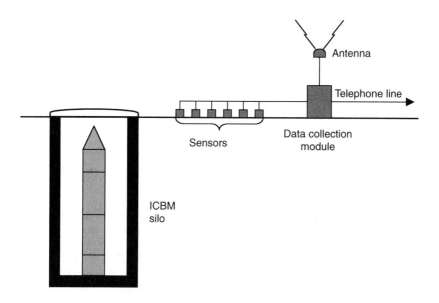

Figure 4.1—Deploying Sensors Outside Silos for Early-Warning Information

If extra confidence is required, the silo door, which must open before the missile is launched, can also be monitored. This can be accomplished by placing one or more seals made of optical fiber over the door so that it cannot be opened without breaking the seal. The data module checks the state of the seal by sending a light pulse down the fiber to see if a unique pattern at the end of the fiber cable has changed. Such seals are widely used by the International Atomic Energy Agency to monitor civil nuclear programs around the world.

Each module samples its sensors several times a minute and transmits the data to a central collection system at each base. The central system collects the data from all modules at the base and transmits those data directly to the monitoring and host countries.

One requirement for any early-warning system is that it be extremely reliable. In addition to using the redundant and complementary sensors described above, the system in this approach would increase redundancy by using two transmission modes that are independent of each other. For example, the transmissions from each base to the monitoring country would be carried by a dedicated satellite line and a dedicated land line. Each module would transmit its data via radio and telephone line to the central facility at the base. The module would get power from the electrical grid and have a battery-based power supply as a backup. The central data facilities would also have backup power for any local power failures.

Early-warning systems must also be able to minimize false alarms: a system that sounds an alarm every time the wind blows, rain falls, or a mouse walks by is not useful. It must also use redundant or complementary sensors as much as possible to reduce the chances that a wayward cow or a squirrel with a taste for optical fiber does not create a crisis. In addition, the system must allow the monitored party to conduct normal operations with only minimal disruptions.

For silos, the basic sensor suite should pose no problems. If fiber optic seals are used to monitor silo doors, the system must allow for routine maintenance on the missile, which in some cases involves opening the silo door to remove warheads, guidance components, or even the entire missile. This can be handled in one of two ways: trust the monitored country to reseal the door with new seals, or have someone from the inspecting country install the seals after maintenance is complete. In either case, any maintenance that is to take place should be announced in advance to the inspecting country. The second approach provides a higher level of system confidence but is more expensive, since it requires inspectors to have a regular presence at each base. The first approach may be acceptable, though, since the seismic and other sensors would remain, and sensors on the other silos would indicate that no large launch had occurred. In addition, door latch sensors could be added to ensure that the silo door is closed after maintenance is complete.

If the system for monitoring launches of silo-based ICBMs works well, it could be expanded to other platforms. The easiest application is to monitor nuclear weapons storage facilities for bombers. Mobile missiles and submarines are much more difficult to monitor because of the need to ensure their stealthiness and mobility, which are key to their survivability. It may be possible to devise schemes that could do this, but it would require extensive cooperation and very intrusive measures. Mobile missiles in garrison would be easier to monitor (Option 8 examines this issue in detail). As long as Russia and the

United States continue to believe that silo-based ICBMs are central to any first strike, however, the system may not need to be expanded beyond silos.

Evaluation

Contribution to Reducing the Risk of Nuclear Use. A cooperative early-warning system based on the Kurchatov proposal applied to silos will help reduce the chances of accidental nuclear use by improving the quality of early-warning information available to Russia and the United States. It also increases decision time and creates a new avenue for cooperation between the militaries in both countries. This option offers several improvements over Russia's existing system for a relatively low price. It could even improve the timeliness and quality of the U.S. early-warning system. The Kurchatov approach could complement the early-warning data that the United States and Russia will share in the Joint Data Exchange Center in Moscow by providing confirmation that the shared data are accurate and not being manipulated. Like other approaches to improving early warning, this option has no effect on the operations of either country's forces.

By itself, however, this approach does not provide Russia coverage of U.S. submarine launches or bombers (although the concept could possibly be expanded to include these platforms). Furthermore, it addresses only U.S. and Russian forces, not other potential sources of missile launches or false alarms around the world. Only a global, space-based system can provide such comprehensive coverage.

For Russia, the silo-monitoring system could add as much as 15 to 20 minutes of warning time during the roughly 17 hours each day that experts have estimated Russia cannot view U.S. missile fields (see Table 4.3, above). If, as discussed earlier, Russia has lost its remaining satellites because of a fire in its ground control center and thus has an early-warning system based solely on ground-based radars, the cooperative early-warning system could provide the extra 15 to 20 minutes of warning 24 hours per day.

The United States could also benefit from using such a system as a complement to its existing, well-functioning satellite and radar early-warning system. By detecting a launch before the missile rises above the clouds, the system would provide the United States with a few additional minutes of warning. And by providing another method for confirming an attack, it would improve the quality of U.S. early-warning information and help resolve false alarms. In addition, the United States could use the system to help calibrate the new SBIRS-High early-warning satellites that it plans to deploy in this decade.

For both the United States and Russia, the fact that each missile is monitored individually and by several sensors makes the system inherently robust against false alarms (such as errant messages from the occasional sensor)—perhaps even more robust than space-based sensors. Such detailed information about each silo means that indications of a launch can be viewed in the context of the entire silo-based force.

Effect on Current U.S. Strategies and Targeting Plans. No effect.

Effect on U.S.-Russian Political Relations. The cooperative nature of Option 2 presents real opportunities to improve U.S.-Russian relations. This approach could also help build each country's confidence about the other's intentions. In addition, it would increase the cooperation between the two countries' military establishments by creating a requirement that they interact on a regular basis, and it could provide an important cooperative experience base and test-bed for developing techniques to monitor forces taken off alert. (This approach is explored in Options 7 and 8.)

Effect on Other Major International Actors. No effect.

Effect on Prospects for Achieving Nonproliferation and Counter-terrorism Goals. No effect.

Feasibility and Affordability. One of the appealing aspects of this option is that it builds on the experience and hardware of an ongoing cooperative monitoring experiment. In addition, it is not very intrusive. For these reasons, Option 2 can be implemented fairly quickly, particularly as a pilot project.

This approach could cost significantly less than most of the other approaches we explored for improving Russia's access to reliable early-warning information. Perhaps most important—especially for Russia—is that the cooperative early-warning system would be inexpensive to operate over the years compared to any space-based system. (The notable exception is the Joint Early-Warning Center, which will also be relatively inexpensive. But this center has the disadvantage of requiring each country to trust that the other has not falsified the data.)

One issue the silo-based system would have to address is that it has not proved its effectiveness yet in an early-warning mode. Russia and the United States have devoted a good deal of work to developing and demonstrating the cooperative system for monitoring fissile materials, but it has not been tested for early-warning applications, which place extremely high requirements on accuracy, reliability, and false alarm minimization. An early-warning system that creates spurious indications of launches could be worse than nothing at all.

Finally, if the monitoring approach is expanded to mobile missiles and submarines, it may prove difficult to implement. Tricky issues arise, such as how the modules are to be attached to submarines and how to make the modules on submarines and mobile missiles resistant to tampering. It also raises important questions about how to balance the need for information with the need for submarine and mobile missile survivability. If such an approach is deemed useful, these problems might be addressed by conducting a series of joint U.S.-Russian experiments to test concepts and hardware.

Effect on Incentive to Strike First with Nuclear Weapons. Option 2 improves crisis stability by providing important confidence that ICBMs have not been launched. By providing the launch status of every silo individually, the system actually provides more detailed information about each silo than a space-based system does. It does not, however, provide information about submarines or launches from other countries.

Ability to Monitor or Verify Implementation of the Option and Effect of Cheating. Option 2 provides a good way to monitor implementation and detect cheating as long as the module and sensors are designed correctly (self-protecting and complementary) and tested for vulnerabilities. (Cheating, in this case, means tampering with the silo-based early-warning system.) Also, the system's distributed nature makes it more difficult to spoof or to entirely disable. It would be essential to ensure that the system is not vulnerable to tampering, however, if it is the only system Russia ends up relying on for early-warning information.

OPTION 3: IMMEDIATELY STAND DOWN ALL NUCLEAR FORCES TO BE ELIMINATED UNDER THE 2002 MOSCOW TREATY

This option is a possible solution for two contributing factors:

*   Nuclear forces kept at high day-to-day alert
*   Perceived vulnerability of nuclear forces or command and control systems

Table 4.5 summarizes the evaluation of Option 3.

Background

Option 3 immediately deactivates all forces scheduled for elimination under the 2002 Moscow Treaty—i.e., the Strategic Offensive Reductions Treaty (SORT) signed in May 2002 by Presidents Bush and Putin in Moscow. During their November 2001 summit in Crawford, Texas, the two Presidents agreed to cut their nuclear arsenals by

Table 4.5

Evaluation of Option 3

Nuclear Safety Criterion	Option: Immediately stand down forces to be eliminated under the 2002 Moscow Treaty
Contribution to reducing the risk of nuclear use	Positive
Effect on current U.S. strategies and targeting plans	N/A or negative
Effect on U.S.- Russian political relations	Very positive
Effect on other major international actors	N/A
Effect on prospects for achieving nonproliferation and counterterrorism goals	N/A
Feasibility and affordability	Very positive
Effect on incentive to strike first with nuclear weapons	Positive
Ability to monitor or verify implementation of the option and effect of cheating	N/A or positive

two-thirds. President Bush indicated that he would cut deployed strategic nuclear weapons to between 1,700 and 2,200; President Putin said he was interested in a range between 1,500 and 2,200.

In January 2002, the Bush administration followed up on the Crawford summit by announcing the results of its Nuclear Posture Review. The review, while keeping the goal of 1,700–2,200 operationally deployed warheads, indicated that this force level would not be achieved until 2012. Before 2012, nuclear reductions would be only gradually implemented, as the United States would reduce its forces from around 6,000 strategic warheads to around 3,800 by fiscal year 2007.

Negotiations that began after the Crawford summit culminated in the Moscow Treaty, a brief one-page document that codified much of what the Bush administration presented in the Nuclear Posture Review.

Specifics

The specifics of Option 3 are rather simple. Both the United States and Russia immediately "deactivate" enough nuclear forces to reach the target of having only 1,700 to 2,200 operational deployed warheads. For the United States, these forces would include nuclear forces already scheduled to be deactivated: 50 Peacekeeper missiles, four Trident submarines, and several hundred warheads, which would be removed from both submarines and ICBMs. For Russia, these forces would probably include all 154 SS-18s, most of its SS-19s, all of its SS-24 rail-mobile and silo-based missiles, and all Delta III and older ballistic missile submarines. This option assumes that the warheads from the missiles are removed and placed in central storage. The smaller number of warheads on each missile could be verified using the on-site inspection provisions of the Strategic Arms Reduction Treaty (START) I. Reversing these procedures would take a few weeks.

There are several historical precedents for this kind of unilateral step. In September 1991, the first President Bush ordered that all 450 Minuteman II missiles and the missiles on 10 older Poseidon submarines be taken off alert. These forces were slated to be removed under START I, which at that time had yet to be ratified by the United

States or Russia. This action occurred shortly after the August 1991 coup attempt in Russia. Within a week, Russia announced that it would reciprocate by taking off alert some 500 silo-based missiles and the missiles on six ballistic missile submarines.

Evaluation

Contribution to Reducing the Risk of Nuclear Use. Option 3 reduces nuclear risk in three ways. First, as President Bush has said, the United States has a new and more positive relationship with Russia. In the 2002 Nuclear Posture Review, Secretary of Defense Donald Rumsfeld went further: "As a result of this review, the U.S. will no longer plan, size, or sustain its forces as though Russia presented merely a smaller version of the threat posed by the former Soviet Union."[11] An immediate demonstration of U.S. willingness to rapidly reduce its nuclear forces will help convince a skeptical Russia that the United States is truly interested in a new relationship. Option 3 reinforces this view by rapidly increasing the pace of nuclear reductions. This lowering of the nuclear temperature should help persuade Russia that the United States has no intention of using nuclear weapons against it and could help lay the foundation for further improvements in nuclear safety.

Second, Option 3 reduces the risk of nuclear use because it helps address the poor first-strike stability between U.S. and Russian forces, a situation likely to worsen over the next decade unless steps are taken. According to one source, even without a formal arms control process, Russia's nuclear forces are likely to shrink to 1,100 operational deployed nuclear warheads by 2010 because of economic factors.[12] If the United States continues to maintain around 3,800 such devices until 2007, the imbalance between the two countries will grow. However, if both sides mutually deactivate their forces to between 1,700 and 2,200 warheads immediately, a better balance could be restored.

---

[11]Donald R. Rumsfeld, *Foreword to Nuclear Posture Review Report* (Washington, DC: Department of Defense), January 8, 2002, p. 2.

[12]Podvig, *Russian Strategic Nuclear Forces*, p. 577. Estimate by 2010 is 300 single-warhead silo-based and mobile Topol M missile systems, seven Delta IV submarines with a combined total of 448 nuclear warheads, and a strategic bomber force of 30 Tu-95MS Bear H bombers, which can carry up to 360 long-range cruise launch missiles. Similar estimates have been made by other sources.

Third, by immediately deactivating the forces spelled out in this option, Russia will eliminate some its oldest and most accident-prone forces. The SS-18s and SS-19s are both nearing the end of their service lives and were scheduled to be decommissioned under START II. If Russia decides that it needs to keep its SS-18s and SS-19s to offset a U.S. missile defense system or a larger U.S. nuclear arsenal, it might try to extend their operational lives until 2008. This could be dangerous, because older nuclear forces may be more accident prone than are newer, more modern forces. Option 3 solves this problem by immediately deactivating these forces.

Effect on Current U.S. Strategies and Targeting Plans. This approach's effect on U.S. strategies and targeting plans is only modest. Indeed, the Nuclear Posture Review indicates that the Bush administration expects the number of forces allowed under the treaty will be sufficient by 2012. This option just accelerates that timeline.

However, the Nuclear Posture Review and the 2002 Moscow Treaty indicate that the Bush administration is uncomfortable reaching a level of 1,700 to 2,200 warheads before 2012. This could be because it thinks there is some chance Russia will revert to a more aggressive nationalist foreign policy. But a Russian buildup will likely be preceded by a sharp deterioration in U.S.-Russian relations and could be detected long before it became militarily significant, giving the United States time to change its posture.

Effect on U.S.-Russian Political Relations. Option 3 should have a very positive effect on U.S.-Russian relations. It indicates that the United States is not trying to take advantage of Russia's economic weakness to gain a strategic advantage in nuclear weapons. It can also serve as a symbolic gesture demonstrating that the United States and Russia are no longer strategic adversaries, at least in the field of nuclear weapons. Option 3 is also an immediate step that may open the way for further improvements in both U.S.-Russian relations and nuclear safety.

Effect on Other Major International Actors (China, Europe, etc.). Option 3 will have a limited effect on other international actors. Even with the nuclear reduction outlined here, U.S and Russian nuclear forces will still be much larger than those of any other nation.

Effect on Prospects for Achieving Nonproliferation and Counter-terrorism Goals. No effect.

Feasibility and Affordability. One of the biggest advantages of Option 3 is that it can be undertaken quickly and can save money sooner. The United States and Russia are already planning to either eliminate or move into reserve many of the forces this option de-activates. Option 3 merely speeds up the process and starts saving money 10 years earlier than now planned.

Effect on Incentive to Strike First with Nuclear Weapons. As noted above, Option 3 should have a positive effect on first-strike stability.

Ability to Monitor or Verify Implementation of the Option and Effect of Cheating. Option 3 can be done with or without verification. The status of nuclear forces could be monitored by the inspection regime under START I, which allows both countries to conduct 10 inspections each year of the front ends of missiles to verify they carry no more than the allowed number of warheads.[13] For bombers, inspectors would check to make sure that nothing is stored in the nuclear weapons storage facilities at a bomber base and that nuclear-armed long-range cruise missiles are not present. With standard START inspections, it could take a month or more to detect cheating. Rapid, widespread cheating, however, should be detectable by intelligence assets and thus might be detected sooner.

The other approach is to have no inspection regime. The Bush administration has argued that formal arms control treaties and extensive verification regimes take too long to negotiate and implement and are no longer important given the reduced tension between the two countries. Therefore, the administration argues for joint steps taken by both nations without formal treaty arrangements. This also allows maximum flexibility in the event that rapid policy adjustments are required because of changes in the international security environment. However, Option 3 will lose much of its effect on relations if U.S. actions are not made somewhat transparent to Russia.

---

[13]For details, see START I, Section VII of the Protocol on Inspections and Continuous Monitoring Activities and Annex 3 of that Protocol (U.S. Department of State, *The Treaty Between the United States of America and the Union of Soviet Socialist Republics on the Reduction and Limitation of Strategic Offensive Arms (START)*, http://www.state.gov/www/global/arms/starthtm/start/toc.html).

OPTION 4: PULL U.S. STRATEGIC BALLISTIC MISSILE SUBMARINES AWAY FROM RUSSIA

Option 4 is a possible solution for four contributing factors:

- Nuclear forces kept at high day-to-day alert

- Perceived vulnerability of nuclear forces or command and control systems

- Short decision times

- Deterrence doctrine or posture reliant on launch on warning or launch under attack

Table 4.6 summarizes Option 4's evaluation in terms of the eight criteria. Note that Option 4 offers two approaches, both discussed below.

Table 4.6

Evaluation of Option 4

Nuclear Safety Criterion	Option: Pull U.S. strategic ballistic missile submarines away from Russia	
	Approach 1: Pull Tridents away from Russia without monitoring	Approach 2: Pull Tridents away from Russia with monitoring
Contribution to reducing the risk of nuclear use	Very positive	Very positive
Effect on current U.S. strategies and targeting plans	Negative or N/A	Negative or N/A
Effect on U.S.- Russian political relations	Very positive	Very positive
Effect on other major international actors	Negative	Negative
Effect on prospects for achieving nonproliferation and counterterrorism goals	N/A	N/A
Feasibility and affordability	Positive	Negative?
Effect on incentive to strike first with nuclear weapons	Positive	Negative
Ability to monitor or verify implementation of the option and effect of cheating	Negative	Positive

Background

Russia has worried for years about U.S. ballistic missile submarines patrolling close enough to its coasts to hit their targets within as little as 10 to 15 minutes. This capability sharply reduces warning time for the Russians. During the Cold War, Russia was able to keep its ballistic submarines close to U.S. coasts. Now, however, for financial reasons and to protect its ballistic missile submarines from U.S. attack submarines, Russia operates its few operational ballistic missile submarines in bastions near Russian territorial waters.

For good reason, the operational policies of U.S. submarines are a closely held secret. So there is no proof that Trident submarines do or do not patrol within 2,000 to 3,000 kilometers of Russia. The Russians, however, continue to worry about it.

The purpose of keeping these submarines away from their targets is to increase missile flight time to at least 25 minutes and thereby reduce Russian anxiety about a very short-warning attack. The fundamental challenge here centers on the fact that submarines are designed to be extremely stealthy in order to avoid detection, which makes it difficult to verify their presence in or absence from a particular ocean region.

We examined two approaches for keeping ballistic missile submarines away from their targets. The first is simply for the United States to keep its submarines further away from Russia. No reciprocal action on Russia's part is required: Russia could respond in kind if it chose to (although its submarines rarely venture near the United States today), but it need not do so. This approach would be a confidence-building measure without any kind of monitoring, and it could be done easily, without negotiations. Increased confidence can be gained with the second approach, but it introduces a complicated monitoring system whose establishment requires detailed negotiations and Russian agreement.

Specifics

Approach 1: Pull U.S. Tridents away from Russia Without Monitoring. The goal of Approach 1 is simple: reduce Russian anxiety by keeping Trident submarines away from Russia. This action increases missile flight times, giving Russia's early-warning system more time to detect a nuclear event and Russia's leaders more time to determine whether the event was an attack. It can be accomplished either unilaterally by the United States or with an exchange of unilateral declarations, much like Presidents George Bush and Mikhail Gorbachev did when they pledged to remove battlefield nuclear weapons from Europe.

Under this approach, the United States would keep its Trident submarines about 6,000 nautical miles (11,000 kilometers) from Russia, near the edge of their missiles' range. Figure 4.2 shows the rough areas in which U.S. submarines would be allowed to operate. From this range, it would take Trident missiles about 25 minutes to reach their targets. The United States would take this action unilaterally; no specific action is required of Russia. However, Russia could reciprocate by declaring that it will not operate its submarines outside their bastions near Russia. Because this option does not require changes to hardware or operations and does not impose monitoring or verification requirements, it has no effect on costs.

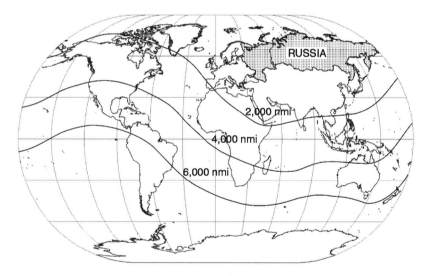

Figure 4.2—Keep-Out Areas for Trident Submarines Under Option 4

An alternative is to keep the submarines completely out of range of their targets. If Trident submarines are carrying D5 missiles loaded with the five W-76 warheads allowed under START II, however, they will have a range of roughly 6,000 nautical miles and will be within range of Russia very shortly after leaving port. To move out of range, they will have to transit to regions of the southern Atlantic and southern Pacific and patrol there (see Figure 4.2), long trips that will burn up more nuclear fuel, an important consideration when it costs about $200 million to refuel a Trident. One solution might be to place ballast on each missile to make its payload as heavy as if it were carrying eight warheads. This would reduce the maximum range of D5 missiles from about 6,000 nautical miles to 4,000, and the submarines would be out of range for several hundred nautical miles after leaving port.

Approach 2: Pull Tridents away from Russia with Monitoring. To increase Russia's confidence that U.S. ballistic missile submarines are staying away from its coasts (and vice versa), Approach 2 adds monitoring provisions to the submarine pull-back described above. This requires that monitoring provisions be negotiated.

The challenge to monitoring a submarine's location is indicating where the submarine is (or is not) without compromising the stealthiness that makes it such a valuable platform for deterrence. The basic concept adopted here is to have the submarine release a special beacon buoy when ordered to do so by the other country.[14] In the case of the United States, the buoys would be launched from the three-inch ejector tubes that submarines use to launch a variety of buoys today. When inspectors ask to verify the location of a specific submarine, the submarine in question releases its buoy. The buoy floats to the surface but does not begin transmitting for 12 hours or longer, giving the submarine enough time to relocate so that the buoy does not endanger its survivability by revealing its location. The buoy transmits a signal that gives its location and a unique code

---

[14]Richard L. Garwin has proposed a variation of this approach (see Richard L. Garwin, "De-alerting of Nuclear Retaliatory Forces," Amaldi Conference, Paris, France, November 20–22, 1997). Another possibility is to use the existing secure communications systems to send a coded message with the submarine's location. To preserve locational secrecy, the transmission could be delayed or an on-board information barrier would just provide a "yes" or "no." The authors thank Frank von Hippel for this insight.

identifying the submarine as the one whose verification was requested. To ensure that the signal is received, it transmits for 24 hours.

For survivability purposes, there would be a limit on how often each country could ask to verify a submarine's position. For our illustration here, we assume that Russia could ask for a submarine's location only once during its 72-day patrol. And to ensure that the force remains invisible, Russia could query only one submarine each week. Thus, if the United States keeps five Tridents at sea, Russia can check for cheating once every two weeks or so, on average. This scheme makes it easier to detect widespread cheating than to detect an isolated incursion by a single submarine.

One way to add a further element of randomness to the inspections is to allow Russia to conduct one or two challenge queries each year in which it can check one submarine more than once on its patrol or two submarines within a week. The United States might also have the right to refuse a request if the submarine in question is being followed by anti-submarine warfare forces. (The numbers for the buoy delay, the numbers of queries per patrol, and the minimum time between queries would all be subject to negotiation. They are used here for illustrative purposes.)

How useful is this scheme? Russia will know that the submarine has been in a particular ocean area and, most important, that it has been far away from Russia's coasts. But it will not be able to pinpoint exactly where the submarine has been, because wind and ocean currents can move the buoy as much as 100 miles from where it is released in the 12 hours before the buoy begins transmitting.[15] Moreover, the submarine can move some 72 to 96 miles during the 12 hours, even traveling at the low speeds of 6 to 8 knots, where it operates most quietly. Together, the submarine and buoy movements create an area of uncertainty 400 nautical miles across. Although Russia may not be able to check another submarine's position for another week, it will know that the first submarine was in compliance at the time of the query and would take several days to move

---

[15]This assumes that winds and currents could move the buoy by as much as 8 knots per hour in an unknown direction, creating an area of uncertainty of 3,000 square miles about the release point.

close enough to significantly shorten its missile flight time. (At 8 knots, a submarine can move only about 200 nautical miles closer each day, or 1,400 nautical miles in one week.) Perhaps more important for reducing the risk of accidental nuclear use, over time this option will increase Russia's confidence in the survivability of its forces.

If a detailed verification regime is used for Option 4, several other issues emerge, perhaps the most important of which is ensuring that the verification agreement includes a way to handle the case of a buoy not functioning properly. One solution is for the submarine to release a second buoy if the first one fails, but this would reveal the submarine's location at two different points. Another solution, one that avoids this problem, is to release two buoys whenever a verification request is made.

Evaluation

Contribution to Reducing the Risk of Nuclear Use. The most significant effect of keeping Tridents away from Russia is that the shortest flight times of U.S. submarine-launched ballistic missiles are increased by 10 to 15 minutes (see Table 4.3, above). This gives Russia more time to assess the accuracy of its early-warning information before it has to decide whether to launch a counterattack, thereby reducing the chances of an accidental launch.

The extra time provided by keeping the Trident force away from Russia could also reduce Russia's general feeling of insecurity with respect to the Trident force. This might induce Russia to take its forces off high day-to-day alert rates, as there would be less concern about the United States launching a debilitating first strike. Of course, the extra time would be wasted if Russia's early-warning system had no ability to view Trident operating areas or if Russia did not believe that the United States had changed its deployment policies.

Effect on Current U.S. Strategies and Targeting Plans. For those who support current U.S. forces and the way they are operated, this option has several disadvantages. One is that keeping submarines far from Russia prohibits them from carrying out some of the missions they have been assigned in the Single Integrated Operation Plan

(SIOP), the U.S. strategic plan for conducting nuclear war. The SIOP creates an orchestrated counterattack in which the timing of each warhead's arrival at its target is critical. Option 4 thus could reduce the U.S. damage limitation capability and allow Russian forces to be launched before Trident missiles had arrived from submarines deployed 6,000 nautical miles away.

Of course, the SIOP could be changed so that the submarines are assigned to targets more appropriate to their new locations. Or the United States could reduce its emphasis on damage limitation in its deterrence strategy.

Another disadvantage is the possibility that the monitoring mechanism set up to assure Russia that U.S. ballistic missile submarines are far away from it could inadvertently reduce the stealthiness of the submarines, making them more vulnerable to attack. One potential concern about the buoy concept is that over time the United States will reveal something about its operational patterns. For example, if U.S. submarines typically leave Kings Bay, float up near England, and then return via Africa, buoys will reveal this pattern and might make it possible for Russian anti-submarine warfare forces to focus their searches in specific areas. This problem can be addressed, however, by doing more to randomize the submarines' paths, something the Navy is already supposed to be doing. Moreover, Russia's anti-submarine warfare capability is extremely limited these days because most of its attack submarines remain in port or have been dismantled.

Effects on U.S.-Russian Political Relations. Option 4 should improve U.S.-Russian relations by reducing Russia's concern that the United States is planning a debilitating first strike against Russian nuclear forces. It gives the Russians more information about how the United States is operating its nuclear forces and therefore more confidence that the United States is not trying to take advantage of Russia's current military weakness. Perhaps most significant, it also provides an important signal that the United States is reducing its emphasis on targeting Russia.

However, an analysis of the implications of Approach 1 or 2 of this option for U.S.-Russian relations reveals a Catch-22 often found in the more intrusive and far-reaching proposals for improving nuclear safety. Approach 1 is a unilateral step the United States can take to indicate that it is not planning a decapitating first strike on Russia. If there were a great deal of trust between the United States and Russia, this step would be enough to reassure the Russians of the U.S. policy and therefore no verification measures would be needed. However, to the extent that suspicion and mistrust remain in the U.S.-Russian relationship, additional verification measures may be needed to reassure the Russians that the United States is actually carrying out this policy.

This leads to Approach 2's complex verification and monitoring scheme, the need for which implies that the U.S.-Russian relationship remains hostile and that, without the scheme, neither country will be confident that the other is carrying out the agreement. However, Approach 2 requires the United States and Russia to reveal at some level how their nuclear submarine forces operate—one of each country's most closely guarded secrets. In a hostile relationship, understanding how the other side's submarine force operates is a great advantage if one wants to attack and destroy those submarines before they can launch their nuclear weapons. Therefore, for either country to agree to a complex verification regime such as that in Approach 2, there will already have to be a great deal of trust in the relationship. But, of course, if there were this much trust, such an agreement would probably not be needed in the first place. A mere announcement of the new policy, such as is suggested in Approach 1, would be sufficient.

Effect on Other Major International Actors (China, Europe, etc.). Option 4 may have an effect on the nuclear deterrents of other major international actors. Both Britain and France have small nuclear arsenals of between 200 and 450 warheads, the vast majority of which are deployed on ballistic missile submarines. In 1998, the British government announced a new policy regarding its nuclear forces, including that the number of boats patrolling at any one time would be reduced from two to one and that the number of warheads on each submarine would be reduced to 48. Interestingly, the British government also announced that its submarines would patrol in a reduced state of alert, with their missiles detargeted, which means they will take days instead of minutes to fire their missiles. According to Britain's Ministry of Defence, no steps were taken to provide transparency as to whether the submarines were operating per the announced changes.

France in recent years has shifted most of its nuclear forces away from land-based intermediate-range missiles and nuclear bombers to its submarine force. When the French ballistic missile submarine force is completed in 2010, it will have four submarines capable of launching nuclear attacks against Russia. French submarines carry the MSBS M4A/B missile, which has a shorter range (6,000 kilometers) than the D5 carried by British and American submarines when they are fully loaded with warheads. Very little information is available about the state of alert or operating scheme of French forces.

How might Option 4 affect British and French ballistic missile submarine forces? During the Cold War, the Soviet Union consistently viewed British and French nuclear forces as part of the much larger U.S. nuclear deterrent. However, because British and French nuclear forces were so much smaller than their U.S. counterparts, Russia was not very concerned about how they were operated. This situation could change, however, if Option 4 were implemented, since it seems likely that Russia, and perhaps domestic public opinion within Europe, will want British and French submarines to operate further away from Russia.

This call for monitoring might be controversial. Ballistic missile submarines represent a much higher percentage of Britain and France's nuclear forces, so they may not be willing to have their forces monitored as suggested in Approach 2. They might also argue that it is less important for their submarines to be monitored because their small nuclear forces cannot mount a debilitating counterforce attack.

However, Britain has already dealerted its ballistic missile submarine forces and may view Option 4 as an important step toward reducing nuclear risk. As for France, it is difficult to gauge how willing it would be to cooperate on changing its submarines' method of operation.

Effect on Prospects for Achieving Nonproliferation and Counterterrorism Goals. No effect.

Feasibility and Affordability. One advantage to Approach 1 is that it can be done quickly, easily, and at very little cost. The President could announce the change and see it implemented within a few weeks. It will impose no changes on the submarines or the way they are operated (other than the parts of the ocean where they patrol). Nor will it require any negotiations or intrusive inspection or verification measures. Finally, Approach 1 will not increase submarine vulnerability; there will still be huge areas of the ocean where they could hide.

Approach 2 offers most of the benefits of Approach 1 and, in addition, makes it possible for each country to monitor that submarines are kept away from their targets. In short, it provides Russia with much better confidence that the United States has pulled its submarines back. By including intrusive monitoring provisions, however, Approach 2 takes more time to implement than Approach 1 does. One solution is to implement a unilateral pull-back, as proposed in Approach 1, and follow it with negotiations to develop a verification agreement. The United States and Russia could also conduct joint experiments on just one submarine to test the monitoring concept before they negotiate a verification agreement.

Effect on Incentive to Strike First with Nuclear Weapons. Calculating the effect of this option on first-strike stability is complex and depends on one's view of whether the option (particularly Approach 2) makes Trident submarines more vulnerable to attack.

If the option has no effect on the survivability of Trident submarines, it provides a slight improvement in first-strike stability. This is because Russian forces will be somewhat more survivable given that they have additional time (15 to 20 minutes) to disperse before a strike by Trident submarines could occur. This option could also increase stability in another, perhaps more important, way: Russia may feel less threatened if it believes that Tridents are no longer lurking close by. However, this option does not change the overall balance (or imbalance) between U.S. and Russian nuclear forces.

If Approach 2 is implemented in a way that does make U.S. Trident submarines more vulnerable to attack, first-strike stability could be significantly undermined. In this case, the United States might have an incentive to launch its missiles quickly in order to use them before they are destroyed by Russian attack submarines. The Russians might have the same incentive, only in reverse, because they would now have a greater incentive to attempt a debilitating first strike if they could eliminate the Trident submarines, the most survivable piece of the U.S. arsenal.

Ability to Monitor or Verify Implementation of the Option and Effect of Cheating. One of the major disadvantages of Approach 1 is that it provides no way for Russia to verify that U.S. submarines are being kept at least 6,000 nautical miles away from its coasts. U.S. submarines are very stealthy and are rarely detected by Russian anti-submarine forces, so it would probably take many months for Russia to detect widespread cheating. Given that there is no verification, it is difficult to argue that this approach will help reduce Russian anxiety about surprise attacks by Tridents. There may be some merit to adopting Approach 1, however, if it serves to demonstrate that the United States is placing less emphasis on nuclear weapons in its relations with Russia.

Approach 2 is an attempt to deal with the verification problem by allowing each country to monitor whether the other's submarines are actually staying away from their targets. This proposal could be effective, but it introduces its own complications and difficulties, particularly the concern that it could compromise submarine survivability. Nevertheless, submarine monitoring should not be rejected out of hand in today's environment. Russia's anti-submarine warfare capability is virtually nonexistent today, and tensions between the two countries are low. Moreover, nuclear safety concerns may be considered more important today than the risk that Russia may gain indirect information about U.S. operational doctrine from a well-designed monitoring system.

OPTION 5: KEEP U.S. ATTACK SUBMARINES AWAY FROM RUSSIA

This option is a possible solution for four contributing factors:

- Nuclear forces kept at high day-to-day launch readiness
- Perceived vulnerability of nuclear forces or command and control systems

- Short decision times

- Deterrence doctrine or posture reliant on launch on warning or launch under attack

Table 4.7 summarizes the evaluation of this option.

Background

Another step the United States could take to reduce Russian insecurities about the survivability of its nuclear forces is to keep U.S. attack submarines out of the Barents Sea and the Sea of Okhotsk, where Russia operates its ballistic missile submarines. The United States routinely deploys attack submarines in these regions to trail Russia's naval forces, including its ballistic missile submarines, and to collect intelligence. In the most recent public incident, at least one U.S. submarine was monitoring a Russian naval exercise in the Barents Sea when the Russian submarine Kursk exploded and sank. Russia has complained about U.S. operations in the past. Since Russian ballistic missile submarines can be tracked by U.S. attack submarines, Russia keeps them in the seas near its coasts where they can be protected by other ships and submarines.

Table 4.7

Evaluation of Option 5

Nuclear Safety Criterion	Option: Keep U.S. attack submarines away from Russia
Contribution to reducing the risk of nuclear use	Positive
Effect on current U.S. strategies and targeting plans	N/A or negative
Effect on U.S.- Russian political relations	Positive
Effect on other major international actors	N/A
Effect on prospects for achieving nonproliferation and counterterrorism goals	N/A
Feasibility and affordability	Very positive
Effect on incentive to strike first with nuclear weapons	Positive?
Ability to monitor or verify implementation of the option and effect of cheating	Negative

Some Western analysts believe that the aggressive posture of U.S. attack submarines is dangerous and unnecessary.[16] They argue that keeping attack submarines away from Russia will improve Russia's confidence in the survivability of its submarines, thereby reducing the pressure on Russia to launch its nuclear forces before they are destroyed. But U.S. submarines operating in Russia's coastal waters perform other functions as well, such as collecting various types of intelligence—a mission that some people believe the United States should not give up entirely. Moreover, since Russia can operate attack submarines near U.S. ports for the same purposes, it could reciprocate by keeping its attack submarines away from the United States.

Specifics

The specifics of Option 5 are very similar to those of Option 4: the United States unilaterally states that it is halting all patrols of its attack submarines in the Barents Sea and the Sea of Okhotsk, where Russian ballistic missile submarines generally operate. Unlike Option 4, however, no monitoring mechanism will be used to verify the location of U.S. attack submarines. Instead, Russia should gain confidence over time that the United States is keeping its attack submarines away from the designated areas as evidence mounts that U.S. submarines are not being detected while tracking Russian submarines or collecting intelligence.

Evaluation

Contribution to Reducing the Risk of Nuclear Use. By keeping U.S. attack submarines away from Russia's ballistic missile submarine bastion areas in seas near Russian territory, Option 5 will have a significant effect on decision time. It gives Russia some confidence that the United States will not try to destroy the Russian retaliatory force, which gives Russia additional time, perhaps hours or even days, to assess the accuracy of its early-warning information before launching a counterattack. This significantly increases decision time

---

[16]See, for example, Jon Wolfsthal, "Kursk: Cold War Causality," *Christian Science Monitor*, August 28, 2000.

relative to other options (see Table 4.3, above). The result is a significantly reduced chance of a Russian launch based on bad information.

The extra time provided by keeping the attack submarines away from Russia can also reduce Russia's general feelings of insecurity about its ballistic missile submarine force. This might induce the Russians to take their forces off high alert during day-to-day conditions, since there would be less concern that the United States could launch a debilitating first strike.

Effect on Current U.S. Strategies and Targeting Plans. For those who support current U.S. forces and the way they are operated, this option has two major disadvantages. First, keeping attack submarines away from Russian coasts will reduce their capability to carry out one of their key missions—destroying Russian ballistic missile submarines before they have a chance to launch their missiles. For those who believe that the damage limitation mission is necessary to deter Russia, Option 5 presents problems, because Russian forces could be launched before U.S. attack submarines could arrive to attack the Russian ballistic missile fleet.

Second, attack submarines perform other functions, such as collecting various types of intelligence. The United States continues to collect intelligence on Russia, and attack submarines, because of their stealthy nature, are a key part of this mission. By agreeing to stop operating submarines near Russia's coasts, the United States will limit its ability to collect this information.

Effect on U.S.-Russian Political Relations. This option should improve U.S.-Russian relations by reducing Russia's concern that the United States is planning a disarming first strike against its nuclear forces. Over time, it could give the Russians more information about how the United States is operating its forces and therefore more confidence that the United States is not trying to take advantage of Russia's current military weakness. It also removes a long-standing source of tension between the two nations, which was highlighted by the initial Russian reaction to the Kursk accident: the sinking was said to be caused by the Kursk colliding with a U.S. submarine.

Effect on Other Major International Actors (China, Europe, etc.). No effect.

Effect on Prospects for Achieving Nonproliferation and Counterterrorism Goals. No effect.

Feasibility and Affordability. Option 5 can be implemented quickly, easily, and at very little cost. The President could announce the change and see it implemented within a few weeks. It will impose no changes on the submarines or the way they operate (other than restricting them from patrolling in the Russian ballistic missile submarine bastions). Nor will it require negotiations or intrusive inspection or verification measures. Finally, Option 5 would not increase submarine vulnerability.

Effect on Incentive to Strike First with Nuclear Weapons. This option should improve first-strike stability, but only if Russia believes that U.S. submarines are staying out of its bastions. It will increase the survivability of Russian submarines in two ways. First, they will no longer be in danger of attack by U.S. attack submarines within Russian coastal areas. Second, Russia had such severe economic problems in the 1990s that it reduced the number of ballistic missile submarines it kept at sea. It now keeps its ballistic missile submarines in port, where they are vulnerable to attack, and reportedly compensates for this vulnerability by keeping some of them on high alert. This means they are prepared to launch their missiles within a matter of minutes without leaving the wharf, making them capable of completing their missions prior to the arrival of a U.S. missile strike. With Option 5, Russian submarines will be more survivable when operating in coastal areas. This may encourage Russia to put additional submarines at sea and to reduce or cease the practice of maintaining ballistic missile submarines in port on high alert.

Ability to Monitor or Verify Implementation of the Option and Effect of Cheating. This option provides no way for Russia to directly verify whether the United States is halting its patrols in Russian coastal areas. However, over time Russia will gain confidence that the United States is keeping its attack submarines away from the designated areas, because evidence will mount that the United States is not tracking Russian submarines in these areas.

OPTION 6: REMOVE W-88 WARHEADS FROM TRIDENT MISSILES

Option 6 is a possible solution for three contributing factors:

- Perceived vulnerability of nuclear forces or command and control systems
- Short decision times
- Deterrence doctrine or posture reliant on launch on warning or launch under attack

Table 4.8 summarizes the evaluation of Option 6 in terms of the eight nuclear safety criteria.

Background

Option 4, above, offers one way to address Russia's concern about the Trident force's ability to quickly attack Russia's forces and command and control systems: keep Trident submarines far away from Russia. Option 6 offers another way: remove the Trident missile warheads that can destroy hardened ICBM silos and command bunkers.

Table 4.8

Evaluation of Option 6

Nuclear Safety Criterion	Option: Remove W-88 warheads from Trident missiles
Contribution to reducing the risk of nuclear use	Positive
Effect on current U.S. strategies and targeting plans	Negative?
Effect on U.S.- Russian political relations	Positive
Effect on other major international actors	N/A
Effect on prospects for achieving nonproliferation and counterterrorism goals	N/A
Feasibility and affordability	Positive
Effect on incentive to strike first with nuclear weapons	Positive
Ability to monitor or verify implementation of the option and effect of cheating	Positive

Trident submarines carry D5 missiles, which can deliver either powerful W-88 warheads or less-powerful W-76 warheads. Although each D5 missile can carry up to eight W-88 warheads, the total number that the Navy can deploy is limited because the United States built only about 400. The combination of the W-88's power (reported to be 475 kilotons) and the D5's accuracy (reported to be less than 100 meters when combined with the Mark 5 reentry body that protects the warhead during its fiery reentry into the atmosphere) is what makes this weapon so potent.[17] According to standard calculations of lethality, two W-88 warheads delivered by D5 missiles have at least an 80 percent probability of destroying an extremely hard target, one that can withstand a blast of 5,000 pounds per square inch.[18] (The comparable figure for an attack by two W-76 warheads delivered by D5s is only about 50 percent.) Such high lethality is equaled in U.S. forces only by the Peacekeeper missile, which is scheduled to be retired in 2003. Along with being lethal, the D5 can be launched close to Russia and can strike its targets within 10 minutes, which places a great deal of pressure on Russia's dilapidated command and control systems.

Specifics

To reduce Russia's concern about surprise Trident submarine attacks on important hardened targets, Option 6 reduces the Trident's explosive power by removing and placing in storage the W-88 warheads deployed on its D5 missiles. The W-88s will be replaced by the smaller, less accurate W-76s, reducing the missile's ability to destroy a hardened command center or silo by nearly half.

Since the United States does not want to reveal sensitive information about its reentry bodies and warheads, Russian inspectors will not be able to observe the warheads being removed from the missiles or in storage. Instead, to confirm that the W-88 warheads have been removed, Russian inspectors can look at the front ends of D5 missiles in the same manner they routinely do under START to confirm that the number of warheads on a missile does not exceed treaty limits.

---

[17]Congressional Budget Office, *The START Treaty and Beyond* (Washington, DC: CBO), October 1991, p. 148.

[18]Ibid.

During the START inspections, the United States removes the missile's nose cone and puts a hard plastic cover over the missile's front end to hide sensitive details about the warheads and the final missile stage, called the bus. The cover has a number of bumps on it corresponding to the maximum number of warheads allowed for that type of missile, eight under START I and five under START II. The covers are designed so that inspectors can see that there are no more warheads than the allowed number, but inspectors are not shown the actual number or the details of the warhead shape. For START inspections, the U.S. Navy uses two different covers for D5 missiles, one for missiles that carry W-88s, the other for missiles that carry the smaller W-76s. The two different covers are required because the two types of warheads are mounted on the missile in different places.

In theory, if the United States uses the W-76 cover, Russian inspectors will be able to confirm that no W-88s are on a missile—a simple inspection shows that the geometry is different. However, Russia has complained about the hard cover for years, arguing that more warheads could be hidden beneath it. (The United States has raised similar concerns about some of Russia's covers.) Furthermore, the design of the covers does not permit the sizes of the two warheads to be compared. The Navy is unwilling to change the covers because it contends that they must be hard to protect sensitive details about the missiles and reentry bodies. Given this debate, Russia is very unlikely to agree that use of the W-76 cover will confirm that the W-88s are gone.

However, since warhead removal will be done unilaterally by the United States expressly to reduce Russian anxiety, this option assumes that soft flexible covers (similar to those the Air Force uses during reentry vehicle inspections of its ICBMs) will be used to allow the Russians to confirm the absence of W-88s. Alternatively, the United States can continue to use the hard covers and hope that Russia begins to believe over time that the W-88s have been removed. This latter approach, however, probably would reduce this option's effect as a confidence-building measure.

Evaluation

By removing the W-88s, Option 6 sharply reduces the hard-target counterforce potential of the Trident force. Tridents will no longer be able to destroy very hard targets with high probability. This option will have no other effect on the operations, survivability, or structure of U.S. nuclear forces.

Contribution to Reducing the Risk of Nuclear Use. By removing the hard-target counterforce capability of the Trident force, Option 6 increases the decisionmaking time available to Russia's leaders—a key goal of those who worry about Russia launching an accidental attack. Russia's decision times will be extended from less than 15 minutes to roughly the 30 minutes it takes U.S. ICBMs to reach their targets (see Table 4.3, above). As with Option 4, the extra time provided—in this case, by removing the W-88 warhead—may reduce Russia's general feelings of insecurity with respect to the Trident force, and this reduction may reduce the chance that another factor, such as inadequate early-warning information, will lead to nuclear use.

Effect on Current U.S. Strategies and Targeting Plans. This option represents a sharp shift for the United States, away from maintaining a prompt counterforce capability in its submarine force. Those who believe that the U.S. deterrent is enhanced by the ability to destroy Russian command and control centers, ICBM silos, and other hardened targets quickly will disagree with this approach. It may also raise concerns for those who believe that the W-88 is an important capability for deterring other countries.

However, this option may not affect U.S. capabilities all that much if, as some have argued, the Trident W-76 warhead has significant potential to destroy hardened targets when delivered by the already accurate D5, particularly if the D5 could be made even more accurate by adding global positioning system (GPS) guidance on its warheads.[19] In that case, the only way to reduce the Tridents' first-strike potential will be to keep them away from Russia or reduce the accuracy of all their warheads.

Effect on U.S.-Russian Political Relations. Option 6 can play a major role in improving U.S.-Russian relations. By removing the W-88 war-

---

[19]See George N. Lewis and Theodore A. Postol, "The Capabilities of Trident Against Russian Silo-Based Missiles: Implications for START III and Beyond," presentation at "The Future of Russian-U.S. Strategic Arms Reductions: START III and Beyond," a meeting held in Cambridge, MA, February 2–6, 1998.

head from its missiles and retiring the Peacekeeper missile, the United States will be removing its two most capable counterforce weapons. This provides an important signal that the United States is moving away from a nuclear strategy based on counterforce, which greatly decreases Russia's reasons for continuing with its launch-on-warning nuclear strategy and will, it is hoped, lead to Russia adopting a more relaxed nuclear posture. Unlike Option 4, Option 6 will be reasonably easy to verify and monitor, something the Russians are likely to consider important in building trust and confidence.

Effect on Other Major International Actors (China, Europe, etc.). Unlike Option 4, this option will have no effect on how British and French ballistic missile submarines are postured.

Effect on Prospects for Achieving Nonproliferation and Counterterrorism Goals. No effect.

Feasibility and Affordability. One major advantage of this option is that it can be done quickly: it is a unilateral measure that requires neither negotiations nor changes to the submarines or the way they are operated. As a result, it does not increase the vulnerability of the Trident force. Furthermore, it is a low-cost option. The only additional expenses would be the Navy's costs to develop and test soft covers for D5s and perhaps some extra funding for an increased number of START inspections each year.

Effect on Incentive to Strike First with Nuclear Weapons. This option increases first-strike stability if Russia's commanders end up being more confident that they and their forces will survive a U.S. first strike long enough to launch a counterstrike. And this, in turn, will make them less likely to launch on warning.

Furthermore, this approach may be able to improve stability without introducing the concerns some have raised about reducing launch readiness or verifying that submarines have been pulled back. It can also be combined with other measures to reduce the risk of accidental nuclear use, including keeping submarines away from their targets, improving early-warning information, and reducing the launch readiness of nuclear forces.

Ability to Monitor or Verify Implementation of the Option and Effect of Cheating. One of the major questions in terms of verifying

Option 6 is how the frequency of Russian inspections of the front end of Trident missiles compares with the time it will take the United States to replace the W-76s with W-88s. Under START, Russia is allowed to conduct up to 10 inspections of ballistic missile warhead loading every year, with no more than two inspections at any one base. Since the United States has only two Trident bases, Russia can view at most four missiles each year, or one every three months, on average. Russia could always conduct another inspection before the three months were up if it were suspicious and had not used up its annual quota at both bases.

Unfortunately, the United States needs only about one day to return W-88s to one missile, and about a week to equip all 24 missiles on one submarine if work continues around the clock. Reloading W-88s on the entire fleet would take longer—from one to two months—because the submarines would have to return from their patrols to be loaded. Submarines can return more quickly, but a sharp change in operational patterns might raise suspicion in Russia and lead to an inspection.

Since the United States presumably is interested in demonstrating that the warheads have been removed, this option makes two changes to START inspection procedures. First, the United States would permit a series of baseline inspections over six months in which Russian inspectors are allowed to look at the front ends of all, or at least a significant portion of, the deployed D5 missiles. Second, the United States would voluntarily increase the number of reentry vehicle inspections allowed each year at a Trident base from two to four, although it could be even more frequent than that.

If the United States were to cheat, it would return its capabilities to current levels. This would not increase Russia's vulnerabilities unless Russia postured its forces so they were more vulnerable than they are today.

OPTION 7: REDUCE DAY-TO-DAY LAUNCH READINESS OF 150 ICBMS IN SILOS

Option 7 is a possible solution for four contributing factors:

*   Nuclear forces kept at high day-to-day launch readiness

- Perceived vulnerability of nuclear forces or command and control systems
- Short decision times
- Deterrence doctrine or posture reliant on launch on warning or launch under attack

Table 4.9 summarizes our evaluation of Option 7. Note that two approaches are presented for this option.

Background

The most straightforward method for increasing decision time is to make it more difficult for both the United States and Russia to launch their nuclear forces quickly. This can be done by immediately remov-

Table 4.9

Evaluation of Option 7

Nuclear Safety Criterion	Option: Reduce day-to-day launch readiness of 150 ICBMs in silos	
	Approach 1: Reduce launch readiness of 150 ICBMs in silos with enhanced START monitoring	Approach 2: Reduce launch readiness of 150 ICBMs in silos with continuous monitoring
Contribution to reducing the risk of nuclear use	Positive	Positive
Effect on current U.S. strategies and targeting plans	Negative	Negative
Effect on U.S.- Russian political relations	Positive	Positive
Effect on other major international actors	Positive	Positive
Effect on prospects for achieving nonproliferation and counterterrorism goals	N/A	N/A
Feasibility and affordability	Positive	Positive
Effect on incentive to strike first with nuclear weapons	N/A	N/A
Ability to monitor or verify implementation of the option and effect of cheating	Negative	Positive

ing nuclear forces from alert, either by shutting the systems off in some way, removing warheads, or, in the case of ballistic missile submarines, moving them out of range of their targets. This approach is sometimes called de-alerting.[20]

The fundamental goal of reducing launch readiness is to increase the time that both countries have available before they must make a decision about retaliating in response to a possible nuclear attack. Ideally, a reduced-launch-readiness force would be completely survivable and require at least a few days or weeks to re-alert in ways that would be very visible to the other country. Because neither country has designed its forces with reduced launch readiness in mind, the challenge is to make today's deployed forces come as close to the ideal as possible.

Historical Experience with Launch Readiness Reduction. There are several historical precedents for reducing launch readiness. In September 1991, the first President Bush unilaterally removed portions of the U.S. nuclear force from alert. He ordered that bombers be taken off strip alert, where they stood loaded with nuclear weapons, ready to take off within 15 minutes. He also ordered that all 450 Minuteman II missiles and the missiles on 10 older Poseidon submarines be taken off alert (these forces were all slated to be removed under START I, which had not yet been ratified by the United States or Russia). This action occurred shortly after the August 1991 coup attempt in Russia. Within a week, Russia announced that it would reciprocate by taking off alert some 500 silo-based missiles and the missiles on six submarines.

In 1994, Presidents Bill Clinton and Boris Yeltsin agreed to stop aiming ballistic missiles at each other's country by programming the missile guidance systems to hit the oceans. This type of action may reduce the damage from an accidental launch, but since the missiles

---

[20]For a full discussion of the case for reducing launch readiness, see Bruce G. Blair, "Dealerting Strategic Nuclear Forces," Chapter 6 in Harold A. Feiveson (ed.), *The Nuclear Turning Point: A Blueprint for Deep Cuts and De-alerting of Nuclear Weapons* (Washington, DC: Brookings Institution), 1999, pp. 101–127; and Bruce Blair, Harold Feiveson, and Frank von Hippel, "Taking Nuclear Weapons Off Hair-Trigger Alert," *Scientific American*, November 1997. For critiques of launch readiness reduction, see Karas, *De-alerting and De-activating Nuclear Weapons*; and Bailey and Barish, "De-alerting of U.S. Nuclear Forces."

can be reprogrammed within seconds, it does nothing to reduce the chances of an unauthorized launch or an intentional launch based on inaccurate or incomplete information.

Two other countries have had some experience with reduced-launch-readiness forces. China keeps its missiles off alert during peacetime—their warheads are removed and their rockets unfueled. And Britain, in response to changes in the world, recently decided to put its entire nuclear force, which is deployed on Trident submarines, on what it calls reduced alert. Although the British government has not described how this is accomplished, it has stated that days rather than minutes will be required to launch missiles.

The Challenge of U.S. and Russian Force Asymmetries. As discussed in Chapter One, U.S. and Russian forces and postures have significant asymmetries that will continue for the foreseeable future. These asymmetries not only create many of the problems analyzed in our study, they also make solutions more difficult. Techniques for reducing launch readiness and verifying compliance are highly dependent on the specific technical and operational details of each system. As a result, the asymmetries can matter.

For example, Russia has based most of its alert forces in silos, the United States on submarines; and changes made to silo-based missiles are easier to verify than those made to submarines. So if there were a proposal to reduce the launch readiness of all U.S. and Russian nuclear forces, verifying the launch readiness of the bulk of the other country's forces would be easier for the United States than for Russia. Russia also operates mobile ICBMs that can drive around the countryside on trucks.[21] And although Russia keeps most of its ballistic missile submarines in port, they can launch their missiles from alongside the pier. By contrast, the United States keeps only about one-third of its submarines in port, but they cannot launch their missiles unless submerged.

There are mechanical differences as well. Russian and U.S. silos are opened by different mechanisms, and the accessibility of batteries

---

[21]Russia also operates roughly 60 SS-24 missiles on railcars. These missiles were supposed to be dismantled under START II, but with the future of that treaty in doubt, it is not clear when they will be eliminated.

and guidance systems—items commonly removed in schemes to reduce launch readiness—often differs in Russian versus U.S. ICBMs. In fact, there are even differences in the ICBMs deployed by one country.

Asymmetries also make it difficult for one solution to affect both countries in a similar and reciprocal fashion. For example, Russia already keeps its few deployed submarines close to Russian waters, where they can be better protected from U.S. attack submarines. So an approach requiring each country to keep its submarines several thousand miles away from the other would affect the United States disproportionately. That does not mean that such an approach should be eliminated, however, since it may be a reasonable approach if the goal is to reduce Russia's anxiety. It just means that easy, symmetric solutions will be elusive. The various asymmetries mean that options for reducing launch readiness must be viewed in the context of a broad strategy for improving nuclear safety when choosing which, if any, to pursue.

Possible Approaches for Reducing Launch Readiness. Because of the enormous complexity of reducing launch readiness and the vast array of possible approaches, we chose to explore two options, one at each end of the spectrum. The first proposes that the United States and Russia reduce the launch readiness of an equal number of ICBMs (150, which is approximately one-third of the U.S. ICBM fleet) through unilateral declarations and without verification procedures. This approach, which is the basis for our Option 7—the option under discussion in this section—is very much like the one Presidents Bush and Gorbachev used when they removed forces from alert in 1991. The second option, at the other end of the spectrum, proposes a complete reduced-launch-readiness regime accompanied by complex, intrusive verification measures. That approach is the basis of Option 8, which is discussed in the next section.

Two commonly discussed methods for reducing the launch readiness of ballistic missiles are (1) remove a key component, such as the batteries or some other component of the guidance system, and (2) remove the actual nuclear warhead. We examined both of these methods, but then chose to investigate the effects of only the first one because we had two major concerns about separating warheads from missiles.

Our first concern was that removing a warhead and placing it in a separate facility could increase the risk of its being stolen, particularly given current worries about the security of Russian warheads. Some Russian officials have expressed similar concerns to us. It is more difficult to steal a warhead from a missile in its silo than from a separate facility, because the heavy silo door has to be opened and the warhead removed from the missile. Our second concern was that the storage facilities containing separated nuclear warheads would make a very inviting first-strike target during a crisis. If one country could secretly reassemble its ballistic missiles, it could easily disarm the other by attacking its nuclear storage facilities, completely eliminating the ability to retaliate. This might even be possible using nonnuclear precision-guided weapons.

Removing a key missile component rather than the warhead limits both of these concerns. Theft of a key nuclear component is, of course, a serious problem, but it is far from the calamity that a stolen nuclear warhead would be. Also, it is much easier to store such things as batteries and guidance systems in a large number of separate facilities, making it far more difficult to destroy a country's ability to retaliate.

Specifics

Option 7 reduces the launch readiness of 150 U.S. ICBMs (one-third of U.S. silo-based missiles) and an equal number (or perhaps an equal proportion) of Russian missiles. This could be a first step toward reducing launch readiness for all forces, because silo-based missiles are the easiest to monitor through START inspections and by satellite.[22] Such an approach has been suggested by several analysts in the United States and Russia, although none has suggested reducing launch readiness for these forces without some verification.[23]

---

[22]If Russia's future force has fewer than 500 silo-based missiles, perhaps a reduction by one-third would be more appropriate. Alternatively, missiles carrying 150 warheads could be subject to the agreement.

[23]See Blair, Feiveson, and von Hippel, "Taking Nuclear Weapons Off Hair-Trigger Alert"; and Alexei Arbatov et al., *De-alerting Russian-U.S. Nuclear Forces: The Path to Lowering the Nuclear Threat* (Moscow: Institute of Global Economic and International Relations), October 2001.

Both countries would disable their silo-based missiles in a way that takes some time to reverse. Silo-based missiles can be removed from alert in various ways, each with its own strengths and weaknesses. The most frequently discussed measures are installing safing pins, disabling the opening mechanism for the silo door, removing the batteries or other components of the guidance system, putting something heavy and large over the silo missile door, and removing the warheads or the entire front section containing the warheads. The details of specific measures may differ for the two countries and even for different types of missiles in one country, because different types of missiles and silos may not be engineered the same way.

This option assumes that the United States will take two steps to reduce the launch readiness of its missiles. First, it will insert a safing pin in each missile. These pins are inserted when the missile is serviced in order to avoid accidents: the missile cannot be launched unless the pin is removed. The United States took its Minuteman II missiles off alert in 1991 over a period of three days by inserting safing pins. The pin is accessed through the silo's service hatch, which is distinct from the silo's main door, through which the missile is fired. Second, the mechanism that opens the main silo door will be disabled by removing the piston that opens the door during a launch. Russia would take similar steps: inserting safing mechanisms and removing the piston that opens the door.

Missile status could be monitored two different ways:[24] Approach 1 uses slightly modified START inspections; Approach 2 uses the same inspections plus baseline inspections and continuous silo monitoring via the silo-based sensors described in the second approach for Option 2 (see Option 2 section, above). Approach 1 is the easier of the two to implement, but inspections could occur too infrequently to catch cheating. Approach 2 addresses this problem by increasing the frequency and thoroughness with which forces are monitored, but it is more difficult to implement.

Approach 1: Monitoring Using Modified START Provisions. Approach 1 monitors the launch readiness of the missiles using an enhanced version of the inspections allowed under START. The treaty

---

[24]Option 8, discussed in the next section, also uses these two approaches for monitoring compliance.

allows both countries to conduct 10 inspections each year of the missiles' front ends to verify that they carry no more than the allowed number of warheads.[25] Only two inspections may be conducted at any one base, however. During an inspection of a silo-based missile, inspectors observe the silo door being opened on the missile that they designate. The treaty requires the missiles to be located not more than 50 meters from the silo for this procedure. The inspectors are then allowed to look down the missile tube to count the warheads, which are usually shrouded so as to protect sensitive information while allowing the warheads to be counted. Russia also shrouds the entire opened silo door, the opening mechanism, and most of the top of the silo. Inspectors get two indications that the door has been disabled. First, they see that the door is opened manually, although this does not prove that the door-opening mechanism has been disabled. Second, this option assumes that the party being inspected will allow inspectors close enough to look at the silo's door-opening mechanism so that they can see that the piston has been removed. This would be an ad hoc confidence-building measure that goes somewhat beyond START protocols. It is not required by the treaty, but it is not prohibited either.

The fact that a missile has been disabled by installing a safing pin cannot be verified under Approach 1 unless inspectors are allowed access to the inside of the missile silo. Likewise, verifying that batteries have been removed from the guidance system can be done only by going into the silo or removing the missile from the silo, both of which require significant changes to the START inspection protocols. Neither of these intrusive measures has been assumed in Approach 1.

Approach 2: Continuous Monitoring. Approach 2 improves the monitoring by using an improved version of Option 2's silo-based sensor system (presented earlier). The primary modification is the use of fiber optic seals on the main silo door and access door to provide continuous monitoring of each silo and confidence that each missile has not been re-alerted. Any time a silo door is opened to reinstall the opening mechanism, the sensors will detect it.

---

[25]For details, see START I, Section VII of the Protocol on Inspections and Continuous Monitoring Activities and Annex 3 of that Protocol (U.S. Department of State, *START*).

Approach 2 also improves confidence that the missile has been disabled in the first place by allowing each country to inspect each silo after its disabling via the modified START inspections discussed above. Since the goal of this option is to increase the confidence of both countries, it seems reasonable that the United States and Russia will both be willing to increase the number of inspections over what START currently allows.

Evaluation

Contribution to Reducing the Risk of Nuclear Use. Reduced launch readiness can reduce the risk of nuclear use in two fundamental ways. First, by taking forces off alert and increasing the number of steps necessary to launch a nuclear weapon, it greatly reduces the chances that rogue forces or terrorists will be able to launch nuclear weapons. For example, a several-hour delay before an ICBM can be launched could be enough to prevent missiles from being launched accidentally, either by a rogue actor or through an error in the command and control system. Option 7 will have only a modest effect on reducing these risks, however, because only 150 silos are affected.

Second, reduced launch readiness can give each country additional time to reach a decision about retaliating in response to a possible nuclear attack. Option 7 will have only a fairly limited effect on decision time, however, since it applies only to one-third of the silo-based missile force. Only a day or so would be needed to re-alert the first missile, and a week or more for all 150 silos, but all the other silos and submarines would continue to operate at current levels of alert.

In sum, Option 7's immediate gains in reducing the risk of nuclear use would be limited. However, its real significance is that it would be a first step in the long, complicated process of reducing the launch readiness of nuclear forces, which, if properly implemented, could significantly reduce the risk of accidental or unauthorized use.

Effect on Current U.S. Strategies and Targeting Plans. Option 7 should have no more than a small effect on current U.S. strategies and targeting plans because it concerns such a small number of forces.

Effect on U.S.-Russian Political Relations. The effect of Option 7 on U.S.-Russian relations is likely to depend in large measure on the details of the option and the state of relations when it is implemented. If relations have improved to the point where standing down nuclear forces becomes an important symbol of progress, the effect on relations is likely to be positive. If the deterrent strategy and mindset have not changed much from those of today, any significant step is likely to be viewed negatively.

Option 7 requires less improvement in relations than the more far-reaching approaches do, since it affects only part of the force, but it still requires some progress from where the two nations stand today.

Effect on Other Major International Actors (China, Europe, etc.). As noted earlier, both Britain and China already have some experience in keeping their forces at reduced levels of launch readiness.

Among the major powers, China has gone the furthest in keeping its small nuclear forces off alert. During peacetime, China reportedly keeps its missiles off alert, their warheads removed and their rockets unfueled. Because of its current nuclear posture, China is likely to react positively to steps by the United States and Russia to reduce the launch readiness of their forces. However, the Chinese are unlikely to join any monitoring scheme that requires them to reveal sensitive information about their nuclear forces. The Chinese have traditionally been unwilling to engage in arms control discussions, their argument being that they are currently at a disadvantage because the United States and Russia have much larger nuclear forces. According to the Chinese, any arms control discussion based on the current global nuclear balance would leave them in a permanently inferior position.

Britain, during its last nuclear review, changed the posture of its forces, announcing that its submarines would take days instead of minutes to fire their missiles. However, no steps were taken to allow anyone to verify or monitor whether the submarines had changed the way they operate. Britain is therefore also likely to react favorably to a U.S.-Russian launch readiness reduction proposal, since it would have little effect on Britain's nuclear posture. It also seems likely that Britain will be willing to incorporate its nuclear forces (including allowing monitoring) if launch readiness reduction is turned into a global proposition. And if the United States, Russia, and Britain were to reduce the launch readiness of their forces, France, with its small nuclear force, may join the effort as well.

Effect on Prospects for Achieving Nonproliferation and Counterterrorism Goals. By taking forces off alert and increasing the number of steps necessary to launch a nuclear weapon, launch readiness reduction could greatly reduce the chances that rogue forces or terrorists will be able to launch nuclear weapons. However, if not done carefully, it could have exactly the opposite effect. If the measures instituted to stand down the forces include removing warheads from missiles, the separated warheads could be stolen if not protected properly.

Specifically, though, Option 7 will not affect U.S. nonproliferation goals, because it will not remove warheads. This choice was made because of concern in the West about the security of separated warheads in Russia.

Feasibility and Affordability. One of the biggest advantages of Option 7 is that it can be implemented relatively quickly and focuses on the forces that are the easiest to verify—silo-based ICBMs. No formal negotiations are required, and the modest deviation from the START I inspection protocol (allowing inspectors to view the mechanism for opening the silo door) can be done informally and unilaterally. Taking advantage of the silo-based early-warning system outlined in Option 2 for continuous monitoring would not add complications beyond those discussed for Option 2.

Effect on Incentive to Strike First with Nuclear Weapons. Because it concerns such a small portion of the U.S. and Russian forces, Option 7 will probably not have a direct effect, positive or negative, on the incentive to strike first. Taking this small part of the ICBM force off alert will not appreciably affect force survivability, and cheating will not increase one country's force by very much.

Ability to Monitor or Verify Implementation of the Option and Effect of Cheating. Option 7 offers two different methods for monitoring: using modified START inspection protocols, and supplementing those protocols with sensors that provide continuous monitoring. In general, START procedures alone will only be able to detect cheating within a month or so. Sensors can provide nearly instantaneous information about cheating, but they are more difficult to negotiate and implement.

Approach 1: Monitoring Using Modified START Provisions. The disabling of 150 silo-based missiles can be monitored by the START on-site inspections mentioned above (with modifications). Each country could check the status of a single silo 10 times each year, or an average of once every five weeks, which allows far more time than a country would need to re-alert its force either by jacking silo doors open or fixing the opening mechanism. If half of the 10 reentry vehicle inspections allowed by START were devoted to missiles on submarines, however, ICBM inspections would take place only every 10 weeks on average.

If cheating were conducted on a broad scale, it might also be detected by photoreconnaissance satellites. Large numbers of open silo doors would be unusual and noticeable from space. So might the activity of large numbers of ground crews working to open the silo doors or repairing the opening mechanisms. Whether such activity could be detected quickly enough would depend on how often the satellites were tasked to look at the missile fields, what the weather might be, and how carefully the activity is disguised. Attempts to disguise cheating would slow down the work process, however, and increase the chances that it would be discovered during an inspection. For example, suppose that the United States decided to cheat and to avoid detection by satellite by opening only two silos a day at each base, fixing the mechanism, and closing the door. This activity might be indistinguishable from normal maintenance activities, but at such a slow rate, it would take the United States 100 days to repair the doors to all 200 of its Minuteman III silos at Malmstrom Missile Complex, and 75 days to repair all 150 silo doors at each base in the F. E. Warren and Minot Missile Complexes. In both cases, the required time is longer than the average time between inspections if five reentry vehicle on-site inspections are conducted on ICBMs each year, which increases the risk of getting caught. Russia could re-alert its missiles more quickly, because it has more bases with fewer missiles deployed at each.

Approach 2: Continuous Monitoring. Early-warning sensors placed on silos as outlined in Option 2 will provide the means to have nearly instantaneous information about the alert status of the missiles. This information would allow each country to target its START reentry vehicle inspections to silos whose opening mechanisms were suspected of having been tampered with. It could also significantly increase confidence that cheating is not occurring.

In sum, the risks of cheating are relatively small for Option 7 because only 150 silos are involved.

OPTION 8: REDUCE DAY-TO-DAY LAUNCH READINESS OF ALL NUCLEAR FORCES

Option 8 is a possible solution for four contributing factors:

- Nuclear forces kept at high day-to-day launch readiness

- Perceived vulnerability of nuclear forces or command and control systems

- Short decision times

- Deterrence doctrine or posture reliant on launch on warning or launch under attack

Table 4.10 summarizes the evaluation of this option.

Background

The background for Option 7, presented in the last section, applies to Option 8, as well.

Specifics

Under Option 8, the United States and Russia reduce the launch readiness of all three legs of their nuclear forces and agree to extensive verification and monitoring provisions. This option combines the measures taken in Options 4 and 5 (keep submarines back), Option 6 (remove W-88 warheads), and Option 7 (reduce launch readiness of silo-based missiles), although, in contrast to Option 7, it affects all silo-based missiles rather than just one-third of them and also removes mobile missiles, bombers, and submarine-based missiles from alert.

Table 4.10

Evaluation of Option 8

Nuclear Safety Criterion	Option: Reduce day-to-day launch readiness of all nuclear forces	
	Approach 1: Reduce launch readiness of all nuclear forces with enhanced START monitoring	Approach 2: Reduce launch readiness of all nuclear forces with continuous monitoring
Contribution to reducing the risk of nuclear use	Positive or negative	Positive or negative
Effect on current U.S. strategies and targeting plans	Negative	Negative
Effect on U.S.- Russian political relations	Positive or negative	Positive or negative
Effect on other major international actors	Positive	Positive
Effect on prospects for achieving nonproliferation and counterterrorism goals	N/A	N/A
Feasibility and affordability	Negative	Negative
Effect on incentive to strike first with nuclear weapons	Positive or negative	Positive or negative
Ability to monitor or verify implementation of the option and effect of cheating	Negative	Positive

Silo-Based Missiles. Option 8 reduces the launch readiness of ICBMs the same way Option 7 does, disabling the silo doors and safing the missile either by installing safing pins or by removing the batteries from the guidance system.

Mobile Missiles. The launch readiness of missiles that Russia has deployed on railcars and trucks will be reduced by removing the batteries from their guidance systems and disabling the mechanisms that raise the missiles to their launch position. Returning the batteries and fixing the erector mechanism should take a few hours for each missile. Russia keeps its 360 SS-25 road-mobile missiles at nine bases, an average of 40 missiles per base. If two crews worked around the clock at each base to re-alert the missiles, it would take about two and one-half days for all 360.

Bombers. Bomber launch readiness will be reduced by removing the nuclear missiles and bombs at their bases, as well as equipment unique to nuclear weapons, and storing both the weapons and the supporting equipment in bunkers at least several hours from the nearest bomber base. At least one week would be needed to return all of the weapons to the bomber bases. This approach is similar to the one taken in START II for the B-1B bomber, a nuclear bomber that has been converted to a nonnuclear role primarily by keeping nuclear weapons at least 100 kilometers from its bases.

Submarines. The launch readiness of submarines will be reduced in two ways. First, each country's submarines will be kept far away from the other country's coasts and, for the United States, their W-88 warheads will be removed (as in Options 4 and 6). Second, both countries' submarines will remain on what the United States refers to as modified alert while deployed at sea. The United States does this for safety reasons every time submarines return to their base from patrol. Sailors on the submarine remove an electronic component in each D5 or C4 missile so that no missile can be accidentally launched while the submarine is in port. This electronic component is returned to the missiles after the submarine leaves port and before it reaches its patrol area. Access to this electronic component is through the access panel on the missile's equipment section. An access hatch in the submarine's missile tubes allows sailors to enter the equipment section while the submarine is at sea.

According to Bruce Blair and several naval experts, sailors need about 90 minutes to install this electronic component in each missile, and submarine crews typically do two missiles at a time.[26] This means that the first two missiles could be readied within 90 minutes and that about 18 hours would be needed to restore all 24 missiles on a Trident to full alert using normal procedures.[27] (This time could be cut in half, though, if enough sailors were assigned to ready four missiles at a time.) It would take roughly a week for the U.S. submarines to sail from areas at least 6,000 nautical miles from Russia to areas within 2,000 nautical miles, and about the same time for all of

---

[26]Bruce G. Blair, *Global Zero Alert for Nuclear Forces* (Washington, DC: Brookings Institution), 1995, pp. 88–89; and interviews with U.S. Navy personnel.

[27]Blair, "Dealerting Strategic Nuclear Forces," p. 116.

the W-88 warheads to be returned to the D5 missiles on two submarines.

The Russian Navy has no equivalent to modified alert that it practices routinely. Indeed, its submarines are designed to launch their
missiles from port. At least some of these missiles are routinely kept
at a high state of readiness, able to launch within minutes.[28] To
provide a level of reduced launch readiness similar to that of U.S.
submarines, this option will require that the Russian Navy remove
batteries or some other component from its SS-N-20 and SS-N-23
missiles, which, like the electronic components in U.S. missiles, can
be installed at sea, if necessary. If such components cannot be installed at sea, a new approach would have to be taken when the
submarines are at sea. This represents a large operational change for
Russia, though. It counts on having some of these missiles ready
to launch in port to compensate for the small number of submarines and warheads that would be at sea and for the vulnerability
of Russia's at-sea submarines to U.S. anti-submarine forces.

The approaches just described will reduce the launch readiness of
the forces if fully implemented, but confidence in the reliability of
these approaches will require additional monitoring. By adding tags
and seals to the existing monitoring provisions of START, both countries can be made more confident that force launch readiness has
been reduced and will remain that way. However, force monitoring is
complex and should be done in a way that increases confidence and
transparency without increasing force vulnerability or encouraging
cheating, particularly during a crisis. Reconciling these two goals can
be very difficult; indeed, it is often the reason why proposals to reduce launch readiness are rejected out of hand. Any provisions for
monitoring forces will require negotiations to establish mutually acceptable procedures.

We examined two monitoring approaches for Option 8, similar in
philosophy to the two monitoring approaches for Option 7.
Approach 1 uses an enhanced version of the current START monitoring scheme that goes further than the changes suggested in
Option 7. Approach 2 augments the enhanced START inspections of

---

[28]Ibid., p. 110.

Approach 1 with additional measures that allow continuous monitoring of nuclear forces. This proposal, made by scientists at the Kurchatov Institute in Moscow, was introduced in the discussion of Option 2 and applied to continuous monitoring of silo-based missiles in Option 7. In Approach 2 for Option 8, this scheme is expanded to continuously monitor all nuclear forces.

Approach 1: Enhanced START Monitoring

Silo-Based Missiles. As in Option 7, the launch readiness status of missiles is monitored by the inspections allowed under START. That is, some of the 10 inspections of missile front ends allowed each year will be used to verify that silo door mechanisms have been removed.

This option goes beyond Option 7 by introducing three significant additions to the monitoring regime. First, it permits baseline inspections in which inspectors can observe that the pistons have been removed from the door opening mechanisms of all silos. After confirming that the pistons have been removed, inspectors place tamper-resistant fiber optic seals on the silo doors and access hatches. Any attempt to open a door breaks a seal. Seals such as this have already been developed in the United States and elsewhere and are widely used in commercial activities and in the monitoring activities of the International Atomic Energy Agency (IAEA). Second, this option allows inspectors to check the status of the seals during regularly scheduled START reentry vehicle inspections. Third, because workers must enter silos periodically to work on their missiles, each country is required to notify the other in advance of any maintenance work that will break a seal. Inspectors can then check these silos during the next reentry vehicle inspection at that base, confirm that the piston has been removed, and reseal the door. Inspectors can also check the seals on a significant fraction (perhaps 20 percent) of the silos at each base, which they are allowed to select. Additional inspections, beyond the two per base per year allowed under START, could be added each year to increase the frequency with which the silo door opening mechanisms are checked, although that is not assumed here.

Mobile Missiles. To monitor mobile missiles, inspectors confirm that the erector mechanisms have been disabled by having their hydraulic components removed and that the batteries have been removed from the missiles. They then seal the missile access doors to the battery compartments. They also place a seal over the hydraulic connection on the mobile missile launchers where the removed parts must be reattached. At the garrisons, inspectors will be able to see that the launching doors in the roofs of the mobile missile garages have been welded shut and will be able to place a seal on them from the inside.

Bombers. For bombers, inspectors check to make sure that nothing is stored in the nuclear weapons storage facility at a bomber base and then seal the door. At bases where the storage facility is being used for conventional weapons, inspectors will only be able to check for the presence of nuclear-armed long-range cruise missiles, as is the case today under START.

Submarines. For submarines, Russian inspectors confirm to the extent possible during reentry vehicle inspections that the W-88 warheads have been removed. But they will not be able to confirm that the launch readiness of the Trident missiles has been reduced. They could confirm that submarines are staying away from the coasts by using the buoy verification system described in Option 4, Approach 2.

Approach 2: Continuous Monitoring. If the START inspections are supplemented with seals, as described above, the measures for reducing launch readiness outlined in Approach 1 will increase the confidence that forces are remaining off alert during peacetime. But the improvements will not address what is perhaps the biggest critique leveled at launch readiness reduction: it reduces stability during a crisis because neither country can be sure that the other has not re-alerted its forces.

Approach 2 tackles this problem by using a continuous monitoring system similar to the one discussed in Option 7 for monitoring the alert status of ICBMs in silos. Although the system must be adapted slightly for launch readiness monitoring, the concept is the same: provide near-real-time information about the status of the forces. However, in this case the concept is expanded to encompass all nuclear forces, not just silo-based missiles. If such a scheme can be applied successfully to launch readiness monitoring, it will make any attempt to re-alert forces much more visible. Most important, it will make the time needed to detect cheating much shorter than the time needed to cheat—the key to stability for any significant effort to reduce launch readiness.

The early-warning system discussed in Option 2 is derived from a cooperative system being developed by the United States (Sandia National Laboratories) and Russia (the Kurchatov Institute and Arzamas-16) for monitoring storage areas for fissile materials. The system consists of a series of modules that collect data from a variety of sensors and transmit those data to a nearby system that collects data from all modules at the facility (see section on Option 2 for details). The data then go to a monitoring center in the host country (the one being monitored) and by satellite or Internet to the country doing the monitoring. The parts of the system that have been developed so far are designed to make tampering difficult. In addition, the monitoring system uses unique authentication codes to prevent the sensors from being tampered with. Components of this system—including the data collection/transmission modules and some newly designed sensors—have been developed and tested in a series of experiments involving Sandia and the two Russian laboratories. The advantage of using this system for Option 8 is that both countries already have some experience with it and with cooperating to design, build, and operate such a system. They have worked with each other to develop and test it, sharing designs, experience, and test data.

Only a few changes are needed to convert the cooperative early-warning system to the mission of monitoring the alert status of forces. Moreover, if the system is already in use on silos as part of Option 2's cooperative early-warning system or Option 7's continuous ICBM readiness monitoring system, both countries will have had several years of experience with it. Data modules would be attached to every silo and bomber weapons storage facility. They would collect data from a suite of sensors, including fiber optic seals, door switches, motion detectors, and video cameras where applicable. The system's primary function will be to monitor the fiber optic seals to ensure they have not been broken or tampered with. The other sensors will act together to reduce the chances that the seals or the data module could be tampered with. They would also provide alternative sources of information about the missiles' alert status to help resolve any ambiguities or false alarms.

Seals and other sensors will be applied to each launcher during baseline inspections. The status of the data modules and the sensor suites will be checked during START inspections. They will also be checked during the routine maintenance visits this option allows to service the monitoring system, change the batteries, and replace defective components. In addition, this option will allow each country to permanently base a few small teams in the other country in order to check the alert status of and reseal the launchers after any maintenance is done on them and to provide emergency repairs. Both countries will have to announce in advance any maintenance activities that require a seal to be broken.

Silo-Based Missiles. For silo-based missiles, the difference between the cooperative early-warning system in Option 2 (see Option 2, details on monitoring silo-based missiles) and the launch-readiness monitoring scheme is slight. The only change is that fiber optic seals will be placed on the main silo doors and the access launch, as discussed in Option 7, Approach 2.

Mobile Missiles. Russia's road- and rail-mobile missiles are tougher to monitor than silos because their value comes from their ability to hide, which makes continuous monitoring impractical. (The United States has no mobile missiles.) The same issue arises with submarines.

When the missiles are in garrison, the data modules collect and transmit data continuously, just like they do for silo-based missiles. Data modules will be installed on the missile transporter (truck or railcar). Infrared sensors that can detect when a missile motor is burning could also be included. And the module might also have logic embedded in it that would allow it to transmit an alarm if it is removed from its launcher or tampered with in other ways. Seals will be placed (1) on the access door to the battery compartment and on the connection where the hydraulic components would be re-attached to the launcher, (2) on the erector mechanism that would break if the missile were erected in preparation for launch, and (3) on the welds that keep the roof door closed in the garages for mobile missiles.

When the missiles are in the field, however, the data modules are switched to a dormant mode: they collect data on the status of the missile every minute or so, but they transmit those data only if any of the three seals on the transporter are broken. Once back in garrison, the data collected during deployment could be downloaded and sent to the monitoring country. The module on a mobile missile will have to transmit its data via communications satellite. To ensure redundancy, two different satellite systems can be used from vendors such as Inmarsat and Iridium. Power will be provided by batteries inside the module.

One of several issues raised by mobile missiles that must be addressed for a launch-readiness monitoring system to be effective is how to make the system tamper resistant without the self-protection afforded by video cameras. One solution is to allow each country to randomly check the status of one mobile missile per week or month to increase the chances that tampering will be detected. Another issue is how to allow maintenance on the missile and launcher without creating false alarms. Since maintenance is mostly done in garrison, the same system used for silo-based missiles could be used for mobiles: prenotification of the activity followed by an on-site inspector who reapplies the seal. That inspector could also check the missiles coming in from the field to make sure the seals and modules have not been tampered with.

Missiles Based on Submarines. Submarines raise many of the same monitoring issues as mobile missiles do: issues concerning how to monitor compliance without compromising stealth. In addition, submarines operate under water, where satellite communications are impossible. The basic approach adopted here is the one adopted for mobile missiles: the sensors deployed on submarines transmit only when the status of the missiles changes, either because the lid (hatch) to a missile tube opens or motion consistent with a missile launch is observed.

In port, inspectors confirm the absence of the missile's electronic components or guidance cans by looking through the missile access doors on the submarine. Then they place seals on the access doors to the missile tube. The status of the seals is monitored by a data module somewhere inside the submarine. That module is connected to the buoy-launching mechanism outside the submarine.

At sea, the buoys are launched only if the seals are broken. The module for the launch-readiness monitoring system allows two missile access doors to be opened on each patrol. The alarm sounds if a third door is opened. This feature is necessary because sailors sometimes have to open the access door during a patrol to service parts of the guidance system. Although this is not a common occurrence, it can sometimes happen once or twice per patrol.

To overcome the problem of transmitting through water, the system launches two buoys to transmit the alarm. (The second buoy broadcasts the same information as the first to improve the system's reliability.) The buoys rise to the surface, begin transmitting immediately, and continue to replay the message for 24 hours or more to ensure the signal is detected. As in the case of mobile missiles, the signal is sent by military or civil satellites. One issue that must be addressed is how to make the modules and sensors able to function in the harsh sea environment and at depths of 1,000 feet or more. The buoys and a tamper-resistant mechanism must also be developed.

To increase the system's resistance to tampering, each country will be allowed to randomly check the status of one submarine per week. This check will also indicate that the submarine has kept away from its targets (see Option 4, Approach 2, for details). In this case, though, the buoy released for the check will not start transmitting for up to 12 hours to ensure that the submarine has time to relocate and thus avoid detection. In addition, an inspector at each submarine base can check the seals when the submarine returns to port. That same inspector can also reseal any tubes that were opened for maintenance or arms control inspections. Whether the system should regularly broadcast the missiles' status when a submarine is in port is an open question. It might not be useful for U.S. Trident submarines: they cannot launch from port and most of them are kept at sea. It would be more useful for Russian submarines, which are rarely at sea and can launch from port. In their case, a supplemental module could be attached to the existing one when the submarine is in port. The supplemental module could then transmit data regularly to a central facility at the base and on to the monitoring country.

Once submarines are back in port, inspectors will have to confirm the absence of guidance components and reseal any tubes that have been opened. They might also need to service the data module or the buoys and could download data collected during the patrol. Accommodating these visits will require a regular presence at submarine bases. Inspectors could be located at or near the base or somewhere else in the region.

Bomber Weapons. Because bombers can be used for both nuclear and nonnuclear missions, this option seals the storage facilities for nuclear weapons after inspectors confirm that they are empty. It also monitors the buildings where the launchers (such as rotary launchers, pylons, and bomb racks) associated with nuclear weapons are kept. The data modules at each storage location rely primarily on fiber optic seals for the doors and possibly on motion detectors and video cameras. In all other respects, the system works the same way as the system for silo-based missiles.

Evaluation

Contribution to Reducing the Risk of Nuclear Use. Reduced launch readiness can reduce the risk of nuclear use in two fundamental ways. First, by taking forces off alert and increasing the number of steps necessary to launch a nuclear weapon, it greatly reduces the chances that rogue forces or terrorists will be able to launch nuclear weapons. For example, a several-hour delay before an ICBM can be launched could be enough to prevent an accidental launch, either by a rogue actor or through an error in the command and control system.

Second, reduced launch readiness can add to the time each country has before it must decide whether to retaliate in response to a possible nuclear attack. It also forecloses the option to launch quickly in response to an erroneous warning of attack—an event that may be increasingly likely if Russia's early-warning system continues to degrade. In addition, the more confident that Russia is that U.S. forces remain off alert, the less likely Russian leaders are to mistake a sounding rocket, or some other benign event, for a Trident missile.

The comprehensive approach in Option 8 can have a significant effect on decision time. A day or so would be needed to stand up the first missiles, and weeks would be needed to bring the entire force to full launch readiness (see Table 4.3, above).

The significance of this added time depends fundamentally on how much confidence each country has that the other is not cheating and how sensitive each country's deterrent is to cheating by the other country. This raises an important dilemma with respect to reducing launch readiness: If either country feels it will be at a disadvantage if the other country cheats *and* is not confident that it can detect cheating soon enough, launch readiness reduction can have the opposite of its intended effect—i.e., it can create intense time pressures to re-alert forces quickly. This dilemma is discussed further below.

Effect on Current U.S. Strategies and Targeting Plans. Option 8 can have a significant effect on current U.S. strategies and targeting plans in two ways. First, if the U.S. deterrence strategy continues to emphasize a tightly choreographed, rapid response to destroy Russian nuclear forces, reducing the launch readiness of the entire force will eliminate that possibility unless the United States is able to re-alert its force during a crisis. If, instead, the United States adopts a deterrence strategy that does not require a rapid response, launch readiness reduction will not affect its ability to meets its deterrence objectives—unless its force is vulnerable (not survivable) if Russia re-alerts its forces surreptitiously. This brings us to the second significant effect that reduced launch readiness could have: If cheating would make forces vulnerable, Option 8 could interfere with current U.S. strategies and targeting plans. If, however, the United States could verify Russian compliance with high confidence and in a timely manner, these concerns would go away.

In our view, U.S. forces would not be that vulnerable if cheating occurred, because most of them would be at sea—unless U.S. submarine vulnerability were increased by the monitoring system. Increased vulnerability is certainly possible, but it is also unlikely, because the United States would not agree to Option 8 or something like it unless it had high confidence that its submarines would remain undetectable. It is also an open question whether Russia would have the wherewithal to exploit a vulnerability if it did emerge.

Russia is more likely than the United States to be vulnerable to cheating, since it will have far fewer survivable forces. As a result, Russia may be less willing to adopt an approach as extensive as Option 8.

Effect on U.S-Russian Political Relations. The effect of Option 8 on U.S.-Russian relations is likely to depend in large measure on the details of the option and the state of relations when it is implemented. If relations have improved to the point where standing down nuclear forces becomes an important symbol of progress, the effect is likely to be positive. If the deterrent strategy and mindset have not changed much from today, any significant step is likely to be viewed negatively. Before Option 8 can be implemented, the relationship will have to undergo significant improvement.

Effect on Other Major International Actors (China, Europe, etc.). Option 8's overall effect on other major international actors is likely to be positive, as discussed above, for Option 7. However, unlike the case of Option 7, the United States and Russia will be unlikely to make across-the-board cuts in launch readiness unless relations with other nuclear powers are very good and nuclear weapons are not a central feature of their relations with one another.[29]

Effect on Prospects for Achieving Nonproliferation and Counterterrorism Goals. Option 8 will not affect U.S. nonproliferation goals, because it will not remove warheads (see discussion of Option 7).

Feasibility and Affordability. It will be very difficult to negotiate and implement Option 8's complete stand down of all nuclear forces because of the complicated and central security issues involved. In addition, the technical details of the measure's implementation are very complicated because of the differences in the two countries' systems. And modifying the START inspection protocols to monitor compliance could also be challenging. However, Approach 2's continuous monitoring system for all nuclear forces will be much more difficult to negotiate and implement, particularly because each country will want to ensure that it does not undermine the stealthiness of its survivable platforms—its submarines and mobile missiles. The continuous monitoring systems will also be fairly expensive, primarily because they require the continuous presence of inspectors, dedicated telephone lines, and monitoring hardware. The costs

---

[29]For a discussion of how other countries affect U.S. and Russian choices about their nuclear forces, see Roger Molander, David Mosher, and Lowell Schwartz, *Nuclear Weapons and the Future of Strategic Warfare* (Santa Monica, CA: RAND), MR-1420-OSD, 2002 (limited distribution; not for public release).

will also be higher than those for the silo-based early-warning system because of the extra on-site inspections. Costs for this approach have to include those for hardware (modules, sensors, buoys, replacement tags), central collection centers, transmission lines (dedicated communications and Internet lines, including satellite access for mobile missiles and submarines), a monitoring center at the national command authority, on-site inspectors, and initial installation costs. Money will also be needed to develop some of the required hardware.

Effect on Incentive to Strike First with Nuclear Weapons. Whether an option to reduce launch readiness will affect the incentive to strike first depends fundamentally on the confidence each country has in the survivability of its forces and in the compliance of the other country. Option 8 can have a significant effect, either positive or negative. It will increase stability, provided that both countries feel confident that they can detect cheating before it becomes large enough to be significant or that their forces are survivable enough and their deterrence doctrine robust enough to make them relatively insensitive to cheating. Ideally, both conditions would hold.

The asymmetries in U.S. and Russian forces do not readily lend themselves to the ideal, however, particularly for Russia, whose forces are not very survivable today. The current mindset in both countries is such that the threat from cheating will be seen as significant, which means the monitoring system would bear the heavy burden of having to detect cheating quickly. For this reason, a continuous monitoring system such as the one outlined in Approach 2 is probably essential for success. The question is whether such a system can be devised that will provide enough confidence in compliance without reducing confidence in the survivability of submarines and mobile missiles. This question can be answered only by careful research and testing and cooperation between the two countries to develop such a system. Taking smaller steps initially, such as the ones proposed in Option 7, will also be helpful.

If none of the conditions—survivable forces, high confidence in compliance, a deterrence strategy that is not time sensitive—is met, even for one country, the standing down of all nuclear forces could sharply increase the incentive to strike first during a crisis. Uncertainty about compliance may actually increase the chance of nuclear use in a crisis because one or both countries would not feel confident that the other has not secretly re-alerted its forces. If one country responded to the uncertainty by starting to re-alert its forces, that action might be perceived by the other as preparation for a first strike, at which point it might feel pressure to launch a first strike of its own. Under these conditions, reduced launch readiness would create a time dynamic for nuclear forces that could increase rather than lessen the risk during a crisis.

A number of U.S. critics of reduced launch readiness have raised such concerns,[30] as have a number of Russian analysts.[31] The Russian argument, which like the U.S. argument, focuses on the consequences of reduced launch readiness for Russia, goes like this: Russia is already vulnerable to a U.S. first strike. Its silos and command and control system are vulnerable to attack by U.S. ICBMs and Tridents. Moreover, Russia has few survivable forces, because all but a few of its submarines and mobile missiles are kept in port or garrison, where they can be easily destroyed. These forces are not only vulnerable, they can also be destroyed very quickly by Tridents close to Russian shores. Added to this bad situation are two other factors. The United States can also attack Russia's hardened targets with its growing arsenal of nonnuclear precision-guided weapons. Moreover, NATO expansion has brought Russian targets within range of tactical aircraft that can carry nuclear and precision-guided weapons. Russia's only solution is to be able to launch its forces quickly before they can be destroyed. So why would Russia accept reduced launch readiness and foreclose the only option it has?

And yet, Russians are not unanimous on these views. Indeed, several Russian analysts have advocated reduced launch readiness for at least some nuclear forces.[32] But the concerns that Russian critics raise indicate that reduced launch readiness may be as tough to sell (at least in the near term) to Russia's defense establishment as it could be in the United States. Again, the approach taken and the de-

---

[30]See, for example, Bailey and Barish, "De-alerting of U.S. Nuclear Forces."

[31]Evgeny Miasnikov, Center for Arms Control, Energy, and Environmental Studies, Moscow Institute of Physics and Technology, interview on January 28, 1999.

[32]See, for example, Arbatov et al., *De-alerting Russian-U.S. Nuclear Forces.*

tails of its implementation and monitoring will significantly affect how Russia perceives the value of launch readiness reduction.

Ability to Monitor or Verify Implementation of the Option and Effect of Cheating. Option 8 offers two different methods for monitoring: using modified START inspection protocols, and supplementing those inspections by using sensors for continuous monitoring. In general, using START procedures alone will only provide the ability to detect cheating within a month or so. Sensors can provide nearly instantaneous information about cheating, but they are more difficult to negotiate and implement.

Approach 1: Modified START Provisions. Under Approach 1, START monitoring provisions will be significantly modified to allow every silo, mobile missile, bomber and weapons storage facility for bombers, and submarine-based missile to be inspected during baseline inspections to verify that each platform has been disabled as required. Using tags and seals, inspectors will then conduct more annual inspections of each type of platform to confirm they all remain disabled.

Compared to the START provisions used in Option 7, those in this approach will provide a better method for detecting cheating. Cheating on a small scale will still be difficult to detect, but large-scale cheating can be detected within a month or so if inspections are timed and aimed strategically with information from national technical means. If the amount of time is not sufficient because of the far-reaching nature of this option, a continuous monitoring system can be used.

Approach 2: Continuous Monitoring. The continuous monitoring system offers two significant improvements. First, each country will have nearly continuous information about the launch-readiness status of many of the other country's forces. That is, the time required to detect even small-scale cheating will be reduced to minutes, compared to weeks or months if only modified START monitoring is used. (Table 4.3, presented early in this chapter, compares the time to re-alert and detect cheating for Approaches 1 and 2.) These times are considerably shorter than what is needed to re-alert even small portions of the force. Second, this option provides a method for verifying the launch readiness status of at-sea submarines and dispersed mobile missiles. Although it does not provide continuous information on these forces, it continuously monitors their status and releases a beacon to warn that missiles are being returned to alert.

Nevertheless, the addition of continuous monitoring does not increase the time it takes to re-alert forces: the methods for removing forces from alert remain the same and are less difficult to undo than removing warheads or entire guidance systems.

As for the risks of cheating in Option 8, they could be significant, particularly if one or both countries believes that being caught off guard would be a significant disadvantage.

The risks associated with cheating arise mostly in the context of a crisis. That is, in some situations, cheating may make a crisis situation less stable. If both countries agree to reduce the launch readiness of their forces, they have little reason to cheat during normal conditions, because they could get caught. But in a crisis, either country may feel compelled to cheat if it believes its survivable deterrent is inadequate. This brings us again to the point made earlier that the best force for launch readiness reduction is one that is very survivable because it is largely insensitive to cheating. The corollary is that a country with a highly survivable force is less likely to launch a counterattack based on erroneous information, because it can afford the time to make sure the attack is real before responding. In concrete terms, the United States, with most of its warheads deployed on submarines, may worry less about cheating than Russia, with its small survivable force, does.

OPTION 9: INSTALL DESTRUCT-AFTER-LAUNCH (DAL) MECHANISMS ON BALLISTIC MISSILES

Option 9 is a possible solution for four contributing factors:

- Nuclear forces kept at high day-to-day launch readiness

- Short warning times

- Inadequate security and control of nuclear forces and weapons

- Inadequate training precautions

Table 4.11 is a summary of our evaluation.

Background

Option 9 addresses the threat of unauthorized and accidental launches differently than the first eight options do. What distinguishes this option is that it seeks to minimize the consequences of a nuclear launch *after* it has occurred rather than to prevent the launch from occurring.[33]

Option 9 installs a destruct-after-launch (DAL) system that allows each country to destroy its launched missiles in flight via a self-destruct mechanism it has installed. The main criticism of this DAL system is that a country's nuclear deterrent could be rendered useless if a potential enemy were to intercept the codes and destroy the missiles in flight during a real nuclear crisis. The system's advocates argue that this risk can be checked by including relatively simple measures, such as periodically changing the destruct codes or allowing the codes to be entered only during specified time intervals during flight.

Table 4.11

Evaluation of Option 9

Nuclear Safety Criterion	Option: Install DAL mechanisms on ballistic missiles
Contribution to reducing the risk of nuclear use	Very positive
Effect on current U.S. strategies and targeting plans	N/A
Effect on U.S.- Russian political relations	Positive
Effect on other major international actors	N/A
Effect on prospects for achieving nonproliferation and counterterrorism goals	Very positive
Feasibility and affordability	Very negative
Effect on incentive to strike first with nuclear weapons	N/A
Ability to monitor or verify implementation of the option and effect of cheating	N/A

---

[33]The same applies to Option 10, as well. See next section.

There is one historical case of a DAL system used on deployed missiles. The Soviet Union reportedly used a passive DAL system not only on test flights, but also on many of its deployed submarine-launched ballistic missiles and some of its deployed ICBMs. That is, the missile was programmed to self-destruct if it sensed that it was deviating significantly from its intended path; no external commands were required. This system was considered more reliable than the guidance systems or potentially irresponsible range safety officers.[34] By contrast, the DAL system proposed in Option 9 requires an action by decisionmakers to rescind a decision to launch or to reverse an unauthorized or accidental launch.

Specifics

The DAL system discussed here is based on a DAL model proposed by Sherman Frankel,[35] although other approaches have also been proposed.[36] Implementing a DAL system involves creating an operational structure that would be added to current launch procedures.

The first step in activating the DAL system is detection of an accidental or unauthorized launch. Current early-warning systems that use satellites to detect near-infrared radiation from booster exhaust are adequate for use in a DAL system if they are positioned properly. Russia would need to either gain access to an adequate space-based system or develop an early-warning system that adequately surveys its home territory (see Options 1 and 2). The United States would have to make sure that its DSP or SBIRS-High system could observe U.S. as well as Russian launches.

The launch detection mechanisms are continuously monitored by a DAL control center (DALcc). Local DALccs are responsible for monitoring specific missile sites, and if a launch is detected, sending in-

---

[34]Bruce G. Blair, *The Logic of Accidental Nuclear War* (Washington, DC: Brookings Institution), 1993, p. 107.

[35]Sherman Frankel, "Aborting Unauthorized Launches of Nuclear-Armed Ballistic Missiles Through Post-Launch Destruction," *Science and Global Security*, Vol. 2, No. 1, November 1990, pp. 1–20.

[36]Garwin, "Post-START: What Do We Want?"; and Garwin, "De-alerting of Nuclear Retaliatory Forces."

formation about the launch to a national DALcc. The national DALcc then determines whether the launch was intended by proper authorities. If it appears to be accidental or unauthorized, the DALcc notifies personnel authorized to command a launch. Since destruction of a missile after launch reverses the decision to launch, only personnel with the proper authority to command a launch are allowed to make the reversal decision.

When the proper authorities are notified, they make the final determination about whether the launch was accidental or unauthorized. They also determine when it is appropriate to notify relevant countries of the launch. These authorities then have the option to command the DALcc to destroy the missile. If the decision is made to do so, the DALcc sends the appropriate self-destruct code to the missile. An appropriate system of geosynchronous communications satellites is needed to communicate the destruct signal.

Figure 4.3 shows a schematic of this process. Arrows indicate communication and/or transfer of responsibility. The numbers associated with the arrows indicate the sequential steps completed as time elapses.

A mislaunched missile can be destroyed in several ways. The first way is to deactivate the warhead, which involves making complicated alterations to the missile and, since the missile continues along its original trajectory, provides no way for the target country to de-

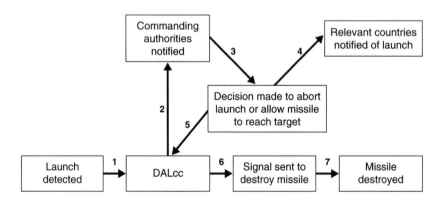

Figure 4.3—Structure of Destruct-After-Launch (DAL) System

termine whether the warhead is, indeed, deactivated. Two other methods—destroying the missile in the boost phase and destroying or disabling the warhead in the midcourse phase—are better ways to build the target country's confidence that the missile or warhead is no longer lethal.

Approach 1: Destroy the Missile in Boost Phase. In Approach 1, the missile is destroyed in the boost phase using conventional explosives. Launch detection by the DALcc can be accomplished using a radar system, and satellites are not needed to communicate the destruct signal. This simplifies the DAL system structure. The technical capabilities needed to fulfill this approach already exist and are employed for test launches. In fact, every U.S. space launcher launched and every ballistic missile launched during a test is equipped with exactly this type of DAL mechanism for safety reasons.[37]

With this approach, the missile is not detected by the target country's radar system, and a crisis can be quickly defused. It allows approximately 3 minutes to communicate a self-destruct signal to an ICBM.[38] This is a short window of opportunity for the DAL system to operate within.

An additional consequence of Approach 1 is that the warhead could land where it would cause considerable damage. Calculations could be performed to determine the ideal point at which to command destruct to minimize consequences, but it would be difficult to do such calculations in the narrow time frames available.

Approach 2: Destroy or Disable Warhead in Midcourse Phase. Approach 2 destroys the threat warhead in midcourse, the roughly 20-minute period when the warhead is coasting outside the atmosphere. This approach can be carried out in one of two ways: (1) using a high-explosive to destroy the reentry vehicle (a method that may not be destructive enough to render the warhead harmless), and (2) by triggering a low-yield nuclear explosion of the warhead itself at

---

[37]Federation of American Scientists, "Submarine Launched Ballistic Missiles, United States Nuclear Forces Guide," http://www.fas.org/nuke/guide/usa/slbm/d-5.htm.

[38]Ballistic Missile Defense Organization, *Harnessing the Power of Technology: The Road to Ballistic Missile Defense from 1983–2007,* September 2000, http://www.acq.osd.mil/bmdo/bmdolink/pdf/power.pdf.

the apogee, using the x-rays emitted to confirm the missile's destruction. In Approach 2, the DAL system has approximately 20 minutes to communicate a self-destruct signal to an ICBM.[39]

Both methods in Approach 2 require extensive changes to the reentry vehicle or the warhead. The first method entails redesigning the reentry vehicle; the second could entail redesigning the detonation mechanism in the warhead. If a low-yield option is not available for the warhead, it may have to be altered, in which case underground testing of the modified nuclear weapons may be required.

To address the concern that an adversary could disable an authorized launch, a digital code will be used to activate the destruct mechanism. This code will be kept by authorities responsible for commanding a launch and the destruction of a launch. If a decision were made to abort a launch, the code would be given to the DALcc and then communicated to a satellite transponder to destroy the missile. Several precautionary steps could be taken to reduce the risk of an adversary intercepting the code. First, the destruct codes could be changed on a periodic basis, making it necessary for an adversary to continuously intercept the code in order to retain the ability to destroy an authorized launch. Second, several codes could be required, and they could be stored at separate points along the chain of command. Lastly, a DAL disable code could be created that would deactivate the DAL system on the missile. Leadership could use this code to guarantee that no one could reverse the decision to launch a missile by using the DAL system.

Another way to diminish the chances of an enemy intercepting the destruct codes and negating a deterrence capability is for each missile to have its own set of codes. An adversary would then have to intercept the destruct codes for every missile that is launched in order to neutralize an entire nuclear arsenal.

Evaluation

Contribution to Reducing the Risk of Nuclear Use. Installing a DAL system can significantly reduce the effects of unauthorized or acci-

---

[39]Ibid.

dental nuclear use. While a DAL system does not prevent the unauthorized or accidental launch of a nuclear weapon, it does provide a process whereby an errantly launched missile can be destroyed. Unlike a missile defense (see Option 10), however, a DAL system would not provide protection from missiles launched by a third country without such a system.

By making it possible to minimize the consequences of a mislaunch, the DAL system addresses to some degree the contributing factors of nuclear forces operating at high day-to-day readiness, inadequate security and control of nuclear forces, and inadequate training precautions. A DAL system also partially addresses the contributing factor of short decision times. It does so by allowing actors a chance to reverse actions that may have been executed in haste, providing 3 minutes for a boost-phase destruct capability and up to 20 minutes for a midcourse capability, although leaders would be unlikely to rely on this additional time.

Effect on Current U.S. Strategies and Targeting Plans. A DAL system has no effect on current U.S. strategies and targeting plans as long as it is secure against misuse. It does not affect a nuclear weapon's ability to reach its target. Instead, it allows a decision to launch a nuclear weapon to be reversed. Given the concern about a DAL system's vulnerability to misuse by an adversary and its potential effect on the U.S. or Russian deterrent, any method adopted must be carefully designed to ensure security.

Effect on U.S.-Russian Political Relations. A DAL system can have a moderately positive effect on U.S.-Russian relations as long as both countries view it as reliable and secure against misuse. It could encourage confidence between the United States and Russia because it diminishes the likelihood of either country suffering the consequences of an accidental or unauthorized launch. In addition, if a joint program were established to develop the communications satellites for such a system, it could foster a positive relationship. A possible drawback may be that a DAL system will not diminish the importance of nuclear weapons in U.S.-Russian relations, instead diminishing the consequences of continued reliance on such weapons.

This option also fulfills the requirement set forth in the 1971 Accidents Agreement. Article 2 of this agreement states that in the event of an accidental or unauthorized launch, "the Party whose nuclear weapon is involved will immediately make every effort to take necessary measures to render harmless or destroy such weapon without its causing damage."[40] Meeting the requirements of this agreement could reenergize the commitment to guard against accidental or unauthorized launches.

Effect on Other Major International Actors (China, Europe, etc.). Option 9 has no significant effect on other major international actors as long as the DAL system is secure against misuse. The only likely effect is a slight reduction in the concern that an accidental or unauthorized U.S. or Russian missile launch could target another major international actor. In the best-case scenario, implementation of a DAL system will have a minutely positive effect. In the worst-case, some international actors, such as Europe and Canada, might be apprehensive about the potential shortfall of the warhead and debris.

Effect on Prospects for Achieving Nonproliferation and Counterterrorism Goals. Counterterrorism goals will be enhanced by a DAL system. September 11, 2001, fundamentally changed the post–Cold War security landscape, verifying that terrorist groups will strive to inflict enormous destruction and will pursue strategic attacks designed to produce high casualties. Access to nuclear weapons will help them achieve these goals. A DAL system will minimize the probability that an unauthorized launch by terrorists will achieve its purpose.

Feasibility and Affordability. The technical capabilities needed to create the DAL system proposed by Approach 1 already exist. It is standard practice to put DAL systems in NASA space launches and DoD missile test launches. Approach 1 is a relatively inexpensive option and can be used as a stepping stone to achieve a more robust system, such as the one outlined in Approach 2. This sequential im-

---

[40]Agreement on measures to reduce the risk of outbreak of nuclear war between the United States of America and the Union of Soviet Socialist Republics. Signed at Washington and entered into force on September 30, 1971. Available at http://www.state.gov/www/global/arms/treaties/accident.html.

plementation of the two approaches may allow the time needed for control centers and procedures to mature.

One challenge in applying such a technology to nuclear arsenals is ensuring an adequate system for both countries. Satellites to monitor home territories will have to be launched, or sensors will have to be deployed on U.S. silos to fill current gaps in Russia's early-warning system. In addition, Approach 2 requires that satellites communicate the destruct signals to the missiles. These could be existing satellites or new ones designed and launched expressly for this purpose. The cost of new satellites could be defrayed to some degree by joint launches of satellites. U.S. and Russian electronic components could be secured and launched on the same satellites.

The key challenge for Approach 2 is implementing the destruct mechanism on reentry vehicles or, if necessary, modifying warheads to trigger a low-yield explosion. For example, can a DAL destroy or damage a reentry vehicle such that the warhead is reliably disabled? Can this destruction be confirmed by either country? By contrast, a low-yield nuclear explosion will reduce or eliminate fatalities on the ground from the launch in a way that can be observed by both countries. But it can also cause extensive damage to satellites operating in low-earth orbits and may be viewed as a precursor attack by the other country.

Effect on Incentive to Strike First with Nuclear Weapons. A DAL system can be seen as encouraging launch on warning, but it is important to remember that the decision to destroy a launched nuclear weapon is not consequence free. Shortfall and debris can cause severe damage. Countries with a DAL system are not more likely to launch a nuclear strike solely because they have the ability to prevent the missiles from reaching their intended targets. Installing a DAL system thus should have no effect on the incentive to strike first.

Ability to Monitor or Verify Implementation of the Option and Effect of Cheating. There is no need to monitor or verify the implementation of a DAL system. The purpose of the system is to provide a way to reverse an accidental or unauthorized launch. Neither country has an incentive to cheat—i.e., to make the other country think it has such an asset when it in fact does not. There is value, however, in demonstrating to the other country that the system is in place.

OPTION 10: DEPLOY LIMITED U.S. MISSILE DEFENSES

Option 10 is a possible solution for three contributing factors:

- Nuclear forces kept at high day-to-day launch readiness
- Inadequate security and control of nuclear forces
- Inadequate training precautions

Table 4.12 summarizes our evaluation of Option 10.

Background

One possible way to reduce the risk of accidental Russian launches is for the United States to deploy missile defenses to protect its territory. Option 10 deploys a limited, ground-based defense aimed at improving nuclear safety.

Two successive U.S. administrations and several Congresses have been committed to deploying national missile defenses, although they have been motivated largely by the desire not to improve nu-

Table 4.12

Evaluation of Option 10

Nuclear Safety Criterion	Option: Deploy limited U.S. missile defenses
Contribution to reducing the risk of nuclear use	Positive?
Effect on current U.S. strategies and targeting plans	Positive
Effect on U.S.- Russian political relations	?
Effect on other major international actors	Negative
Effect on prospects for achieving nonproliferation and counterterrorism goals	Positive?
Feasibility and affordability	Negative?
Effect on incentive to strike first with nuclear weapons	?
Ability to monitor or verify implementation of the option and effect of cheating	Positive

clear safety, but to counter the threat of deliberate attack by emerging missile states. For some advocates, though, protection against accidental or unauthorized launch has been a primary motivation.[41]

In 1988, Senator Sam Nunn delivered a speech to the Arms Control Association calling for a reorientation of President Ronald Reagan's missile defense system known as the Strategic Defense Initiative (SDI). Nunn called for the new SDI program to focus first on developing a "limited system for protecting against accidental and unauthorized missile launches."[42] In 1989, the first Bush administration did reorient the missile defense toward this goal and began developing the Global Protection Against Limited Strikes (GPALS) system, which included ground- and space-based interceptors to protect against accidental or unauthorized launch from the Soviet Union. During the Clinton administration, the possibility of an accidental or unauthorized launch from Russia was regarded as an important rationale for a national missile defense, although it had become secondary to the goal of defending against a limited ICBM threat from rogue nations. For example, the stated goal of the National Missile Defense program in 2000 was "to protect all 50 states from a limited number of long range ballistic missiles launched from a rogue nation or as a result of an accidental or unauthorized launch from a current nuclear power."[43]

Concerns about an accidental launch have continued into the second Bush administration. Representative Curt Weldon, one of Congress's chief advocates for missile defense, said in a May 2, 2001, speech on the House floor: "Today, Madame Speaker, America is totally vulnerable. If an accidental launch occurred of one missile from Russia, from North Korea, which we know now has the long range capability, or from China, we have no capability to respond . . . so the

---

[41]See, for example, Senator Sam Nunn's amendment to the FY97 Defense Authorization Act, "Amendment No. 4180 to the FY97 Defense Authorization Act," Section 1303, National Missile Defense Policy, (b) System Design, "The antiballistic missile system developed under subsection (a) shall—(1) be designed to protect the United States against limited ballistic missile threats, including accidental or unauthorized launches or attacks by Third World countries."

[42]Missile Defense Agency, "Missile Defense Milestones (1944–2000)," http://www.acq.osd.mil/bmdo/bmdolink/html/milstone.html.

[43]Ballistic Missile Defense Organization, "National Missile Defense Integrated Flight Test Three (IFT-3)," BMDO Fact Sheet 110-00-11, November 2000.

first reason we need a missile defense is to protect us against an accidental or deliberate launch." [44]

In determining what type of defense would be most useful against an accidental or unauthorized launch, it is important to decide first what the size and characteristics of such an attack might be. For example, a terrorist or rogue commander who was able to take over one Russian Delta IV submarine and launch its missiles could deliver 16 missiles, each with four independently targetable warheads, for a total of 64 warheads. And these warheads are likely to have sophisticated countermeasures, because they were modernized against a potential U.S. missile defense system in 1988, and Russia may continue to improve their countermeasures to neutralize any defenses the United States might deploy.[45] Another possible threat whose size and characteristics need to be considered is a rogue commander or terrorist taking over one ICBM missile division in Russia. This threat would vary from roughly 20 to 50 missiles that also may carry sophisticated countermeasures.

A national missile defense system could be sized against a smaller threat, perhaps a single missile or a handful of missiles launched accidentally, each carrying one to three warheads. At the moment, the United States has no ability to defend against even a very limited accidental or unauthorized launch, so a missile defense with such limited capabilities (say, 5 to 20 warheads) would be an improvement, as long as the likely threat was very small.

Finally, it is possible that the United States would want a very large national missile defense system to defend it against a large retaliatory strike that might be launched if Russian commanders erroneously believed the United States was attacking.

Specifics

Option 10 deploys a limited, land-based system capable of intercepting a few tens of warheads and whose objective is to improve nuclear

---

[44]Curt Weldon, "Defense of America's Homeland," speech delivered on the floor of the U.S. House of Representatives, May 2, 2001.

[45]Podvig, *Russian Strategic Nuclear Forces*, p. 337.

safety. It would be a midcourse system, which is the most technically feasible approach over the next 10 years or more. Although larger attacks are a possibility, particularly if Russian commanders were to launch a deliberate attack based on erroneous information, we chose this system for two reasons. First, this report repeatedly emphasizes the need for the United States to build a more cooperative relationship with Russia. The Russians could regard a large national missile defense system (such as a space-based boost-phase system) as a threat to their strategic deterrent and thus feel compelled to take steps more apt to lead to an accidental or unauthorized launch. The small missile defense in Option 10 could easily be overcome by even a reduced Russian force.

Second, while the United States has made progress on missile defense technology, particularly hit-to-kill technology, there are still important technical limitations to the kind of system it could build over the next decade or two. Some advocates of a large missile defense want to move immediately to space-based systems. However, an operational hit-to-kill system based in space is at least 15 to 20 years away, and the space-based laser will not be operational until after 2020. We judge these to be too far off to deal with the immediate problems of accidental and unauthorized launch.

The missile defense architecture for Option 10 is based on what was known as Capability 3 under the national missile defense plan that DoD proposed in the late 1990s. There would be a total of 250 ground-based interceptors deployed at two sites, as well as enough radars to track any accidental or unauthorized missile launch that might come from Russia. The interceptors would be based in Alaska and New England, where they would be better situated than a single-site system to protect the United States from Russian ICBM launches over the pole and submarine-launched missiles from the northern Atlantic or Pacific. As many as nine X-band radars would be deployed around the United States and Europe, including in Alaska, Britain, Greenland, North Dakota, and Canada. This system, advertised as capable of intercepting with high probability roughly 20 warheads accompanied by relatively sophisticated countermeasures, might be deployed within a decade or so.

In the past, to deploy a missile defense for accidental launch protection would have required a completely new system, since no system was deployed or under development. Today, with a consensus for deploying a national missile defense, the Bush administration has made deployment of a missile defense with multiple layers one of its highest priorities. It plans to start deploying the first elements of a defense before 2006 and to then continually add to the system every two years as part of a so-called spiral development approach. Therefore, the question for Option 10 becomes, "What changes, if any, would be required to the administration's current plans in order to provide adequate protection against accidental or unauthorized launch from Russia?" Unfortunately, this question is difficult to answer with certainty at this time because the administration has not articulated what final set of systems it plans to deploy. However, this option attempts an answer based on the information that is available today.

Despite the uncertainty surrounding the Bush administration's plans for a missile defense, it is clear that all components of the system proposed under Option 10 would already be under development. For example, the administration is forging ahead with plans to develop the kill vehicle and booster rocket for a ground-based interceptor even if it deploys only a few of them in Alaska and California, as current plans now suggest. In addition, development of the X-band radars, tracking satellites, and command and control systems are proceeding. However, the administration has not articulated plans to deploy more ground-based interceptors than the 10 planned for each of the two test-beds in Alaska. Nor has it indicated plans to deploy any land-based X-band radars. Therefore, in our estimation, Option 10 will require the purchase and deployment of 250 interceptors and as many as nine X-band radars above what is in the administration's current plans. Of course, if the administration's final plans include some of the items listed above, fewer of them would have to be purchased under this option.

Evaluation

Contribution to Reducing the Risk of Nuclear Use. Like Option 9 (installing a DAL system), Option 10 does not prevent the launch of nuclear weapons. Instead, it provides a mechanism to destroy a missile-delivered warhead before it detonates. Its goal is to avert the consequences of an accidental or unauthorized launch after the launch has occurred.

Also like Option 9, Option 10 can address the contributing factors of nuclear forces at high day-to-day readiness and inadequate security and control of nuclear forces by providing a chance to reverse the consequences of an accidental or unauthorized launch. However, because Option 10's missile defense system is limited, it could intercept only a small number of missiles and warheads and could be rendered useless if Russia deploys countermeasures on its missiles that can penetrate the defense.

Therefore, Option 10's effectiveness depends both on the size of any launch and on what steps, if any, Russia takes to deploy countermeasures.

Effect on Current U.S. Strategies and Targeting Plans. The proposed missile defense system will have only a limited effect on current U.S. strategy and targeting plans against Russia. Advocates of missile defense argue that an effective system will enhance the U.S. deterrence posture against emerging missile states, such as North Korea and Iraq, but the U.S. ability to annihilate a rogue state with its massive nuclear arsenal may already be an effective deterrent.

Effect on U.S.-Russian Political Relations. As long as both the United States and Russia view the missile defense presented in this option as limited in size and capabilities, their relations are likely to remain cooperative. The United States may even provide missile defense technology to Russia, thereby allowing it to build its own limited missile defense. However, there could be a negative effect on relations if Russia believes the system is intended to undermine its deterrent. In that case, Russia could become less cooperative on many issues, including those related to nuclear safety.

Effect on Other Major International Actors (China, Europe, etc.). Option 10 could have a major negative effect on international actors, particularly China. At the moment, China has only about 20 single-warhead ICBMs that can reach the United States. Even the relatively small missile defense envisioned in Option 10—with 125 interceptors on each coast—could negate China's deterrent. How China would react to this option is difficult to ascertain. China might have only a mild reaction and merely continue down its established path, building a slightly larger and more modern nuclear force. However, China could also see the system as a U.S. attempt to gain greater flexibility in a U.S.-Chinese regional conflict over Taiwan. In this case, China might substantially increase the size and readiness of its nuclear arsenal, which could lead to a U.S.-Chinese nuclear relationship reminiscent of that between the United States and Russia during their nuclear buildup in the 1960s and early 1970s. The result could be a serious degradation of global nuclear safety.

Effect on Prospects for Achieving Nonproliferation and Counterterrorism Goals. The missile defense system presented in this option might be an effective tool for meeting nonproliferation and counterterrorism goals. Its most important contribution would be to limit the probability of success of an unauthorized launch against the United States. However, if U.S.-Russian or U.S.-Chinese relations deteriorate because of missile defense, U.S. nonproliferation and counterterrorism could be undermined. Russia has been developing technologies to defeat missile defenses since the late 1960s and would likely continue its research in the face of U.S. deployments. China would also be likely to continue its own research efforts. Although these technologies are not likely to spread, one cannot rule out their transfer to states such as Iran and North Korea if relations with the United States were to get bad enough.

Feasibility and Affordability. There are two potential problems with the effectiveness of the midcourse system proposed in this option. First, it may be vulnerable to countermeasures that would undermine its effectiveness even against small numbers of warheads. Second, it could easily be overwhelmed by a large number of missiles, which could not be ruled out in an accidental or unauthorized launch.

There is strong disagreement in the technical community about the feasibility of a midcourse missile defense. Critics of a midcourse system claim it will be susceptible to simple countermeasures. The most complete documentation of this problem is contained in a report by the Union of Concerned Scientists,[46] a report detailing a number of

---

[46]See Andrew Sessler et al., *Countermeasures: A Technical Evaluation of the Operational Effectiveness of the Planned U.S. National Missile Defense System* (Cambridge, MA: Union of Concerned Scientists and MIT Security Studies Program), April 2000.

simple countermeasures that rogue nations could use to confuse the sensors of the national missile defense system, thereby preventing the system from distinguishing between decoys and the warhead. Although there is much speculation about what emerging missile states might be able to do to penetrate the defense, there is no question that Russia's well-developed missile program and 40-year history of developing and deploying countermeasures would enable it to deploy countermeasures that have a good chance of overcoming the defense. Therefore, the midcourse defense posited in this option could be ineffective against even a small accidental or unauthorized launch if Russia routinely deployed countermeasures on its missiles. However, if Russia views the accidental launch protection system as too small to be a threat to its deterrent, it may not deploy countermeasures, and the defense would then have some value in reducing nuclear risk.

Of the 10 options examined in this study, building a missile defense is by far the most expensive, although it presumably would be developed largely for reasons other than nuclear safety. The CBO estimated that Capability 3 of the national missile defense system, which is similar to the system we are suggesting in Option 10, would cost between $51 and $58 billion through 2015.[47] However, this estimate includes all the costs for developing, procuring, and operating the system. As discussed above, it could be argued that the administration is likely to develop most, if not all, of the components as part of its National Missile Defense program. Therefore, the only unique costs for Option 10 would be for procuring and operating the missile defense sites and radars that were for nuclear safety and thus in excess of what the United States would deploy for other purposes. These costs are difficult to quantify at this point, because the administration's plans are unclear. They could range from zero to billions of dollars.

Effect on Incentive to Strike First with Nuclear Weapons. The limited midcourse system in this option—with around 250 interceptors (125 at each site)—might be small enough to have a positive effect on first-strike or crisis stability with respect to Russia, as long as it is

---

[47]Celeste Johnson, "Estimating Costs and Technical Characteristics of Selected National Missile Defense Systems" (Washington, DC: CBO), January 2002, p. 8.

not perceived as a threat to Russia's retaliatory deterrent. If a very small accidental or unauthorized launch occurred, a limited national missile defense system would give the President more flexibility in dealing with the crisis, assuming he had confidence in the system's ability to handle any Russian countermeasures.

However, first-strike stability may decrease if Russia perceives the defense as large enough to counter its survivable, second-strike nuclear forces. This could compel Russia to keep an even greater number of its weapons on high alert so that they could be launched on warning to avoid being destroyed by a U.S. first strike. Whether the defense in Option 10 would be large enough to make Russia feel insecure about its retaliatory deterrent depends on how effective Russia perceives the system to be and how small it believes its survivable deterrent is. The state of U.S.-Russian relations will likely color these perceptions.

Ability to Monitor or Verify Implementation of the Option and Effect of Cheating. Generally, advocates of missile defense do not believe verification is necessary. However, the United States and Russia could, if they chose to, set up an inspection regime to verify the rough capabilities of the missile defense. Because the proposed midcourse system would be based on land, either satellites or on-site inspections could be used to verify how many interceptors the United States or Russia had deployed. This would serve as an upper bound on the system's capabilities, since the system cannot shoot down more warheads than it has interceptors. In fact, the current plans for national missile defense call for several interceptors to engage each incoming warhead in order to give the United States several different opportunities to attack the missile.

Chapter Five

RECOMMENDATION: A PHASED APPROACH FOR IMPROVING NUCLEAR SAFETY AND U.S.-RUSSIAN RELATIONS

Which of the 10 options examined in the previous chapter show the most promise? And how should they be incorporated into a strategy for improving nuclear safety? In our view, a successful strategy for limiting nuclear dangers requires both operational changes in the U.S. and Russian nuclear postures and improvements in the level of trust and cooperation between the two nations. This should be a mutually reinforcing process in which near-term improvements in nuclear safety build confidence and trust between Russia and the United States, thereby enabling more extensive steps in the medium and long term. These dynamics lead us to recommend a phased approach.

Our long-term vision is a U.S.-Russian relationship in which neither country views the other as a nuclear threat. The current relationship between Britain and France is an illustration of this end state. Both are nuclear powers with divergent views on some issues, yet neither would consider using nuclear weapons or even military force against the other to settle a dispute.

Reaching this long-term goal will be difficult, but that is no reason for complacency about nuclear safety issues. As discussed throughout this report, the danger of accidental or unauthorized launch remains a serious problem despite improvements in U.S.-Russian relations. This danger has been heightened by the deterioration of Russia's early-warning system, Russia's economic difficulties, and the continuing U.S. reliance on a damage limitation nuclear strategy. Due to the serious nature of this problem, we recommend that the United States take immediate actions to begin the process of improving nuclear safety as part of what we call a "Nuclear Safety Initiative."

The initiative would begin with a series of unilateral U.S. actions taken to demonstrate both U.S. commitment to reducing nuclear dangers and U.S. interest in a new nuclear relationship with Russia. At the same time, the United States would commit itself to further actions aimed at increasing nuclear safety that would require some time to implement. The hope is that Russia will respond with unilateral actions of its own, but this is not a requirement, at least in the first phase, because the U.S. intention would be to demonstrate that nuclear weapons are diminishing as a factor in its relations with Russia. Unilateral actions would be followed by negotiations between the United States and Russia on further steps that could be taken in the near term to improve nuclear safety. The expectation is that the near-term steps will lead to more-extensive and far-reaching steps to reduce nuclear danger in the medium and long term.

Figure 5.1 graphically illustrates our approach. Starting from today, a series of immediate, unilateral actions are taken to improve nuclear

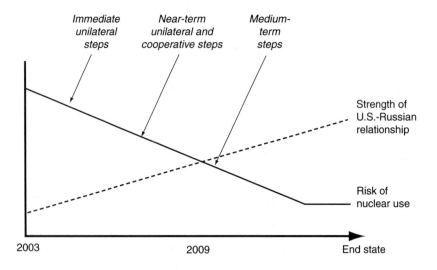

**Figure 5.1—Phased Approach for Improving Nuclear Safety and
U.S.-Russian Relations**

safety. At the same time, the United States commits itself to undertake further near-term actions over the next two to three years that would take additional time to implement. This leads to both a reduction in nuclear risk and a strengthening of U.S.-Russian relations by reinforcing the notion put forth by President Bush that U.S. weapons are no longer intended for Russia. Building on the initial steps, additional actions requiring both nations' consent in the medium term (five to seven years from today) could further reduce nuclear risks and further improve U.S.-Russian relations. If relations continue to evolve in a positive direction in other dimensions as well—including increased trade and cooperation against terrorism—it may be possible to consider additional nuclear safety measures that seem too difficult or risky today.

The timing suggested in Figure 5.1 is notional. From our vantage point today, it appears to be reasonable. But progress could be faster if conditions and leadership allow.

After evaluating all of the options detailed in Chapter Four, we organized the ones we believe to be the most promising according to four criteria: (1) How much will the option contribute to nuclear safety or improved relations? (2) How long will the option take to implement technically? (3) What previous steps (if any) need to be taken before the option can be implemented? (4) How much would U.S.-Russian relations need to improve before both countries felt comfortable implementing the option? Figure 5.2 illustrates our recommended approach for the first three phases of the Nuclear Safety Initiative.

IMMEDIATE STEPS

The United States should take the following four immediate actions to improve nuclear safety: stand down U.S. nuclear forces to the levels specified in the Moscow Treaty, pull Trident ballistic missile submarines away from Russia, pull attack submarines away from Russia, and reduce the launch readiness of one-third of U.S. silo-based missiles. All four actions can be taken quickly and unilaterally by the United States.

At the same time, the United States should commit itself to several other steps that would take some time to implement, either because they are technically difficult or because Russian participation is re-

6 months to 1 year	2 to 3 years	5 to 7 years
Immediate Unilateral Steps	**Near-Term Unilateral Steps**	**Medium-Term Steps**
• Stand down U.S. forces to Moscow Treaty levels	• Eliminate Moscow Treaty forces	• Take equal number of silo-based ICBMs off alert
• Pull SSBNs away from Russia	• Put EW sensors on U.S. silos	• Install sensors on silos to monitor reductions in launch readiness
• Pull U.S. attack subs back	• Remove W-88 warheads	
• Reduce launch readiness of 1/3 of U.S. silo-based missiles	**Begin Consultations on:**	• Adopt a new deterrence strategy
	• Further improving Russian EW systems	• Deploy limited missile defense
Commit to:	• Installing DAL systems	• Deploy DAL system (mid-course)
• Put EW sensors on U.S. silos	• Reducing launch readiness	• Continue negotiations on further steps to reduce launch readiness
• Fund Russian EW radar		
• Continue RAMOS program		

| 2003 | 2006 | 2010 |

NOTE: SSBN = ballistic missile submarine; EW = early warning; DAL = destruct after launch; RAMOS = Russian-American Observational Satellite.

Figure 5.2—Potential Steps for Improving Nuclear Safety, 2003–2010

quired. The most important issue, in our view, is improving Russia's access to reliable, accurate early-warning information. As a result, we believe the United States should commit to putting early-warning sensors on all its silos, funding Russian early-warning radar stations, and continuing the cooperative Russian-American Observational Satellite (RAMOS) program.[1] It is even more critical, in our view, that the United States try to persuade Russia to agree to joint early-warning sensors on silos. This would enhance the unilateral efforts that the United States had committed to in order to provide Russia

_____

[1]After careful consideration, we decided against recommending the launch of additional Oko satellites. While they would provide some additional coverage, their limitations keep them from providing a permanent solution to Russia's early-warning problems.

with greater information about the status of U.S. ICBM forces. The silo-based monitoring system could also become an important test-bed for exploring ways to monitor the launch readiness of nuclear forces.

Taken together, these actions and commitments will send an immediate signal to Russia that the United States is truly interested in nuclear safety and in establishing a new U.S.-Russian relationship in which nuclear weapons do not play an important role.

Our hope is that Russia will respond with unilateral actions of its own, standing down the forces that it plans to eliminate under the Moscow Treaty and promising to keep its ballistic missile and attack submarines away from U.S. coasts. Russia may also respond by becoming open to discussion about improving its early-warning systems.

NEAR-TERM STEPS

The immediate steps could be followed by steps taken over the next two or three years. For these, near-term steps, the United States would fully implement its earlier commitments, such as eliminating all forces it had stood down, placing sensors on all of its silos, having Congress approve money for improving Russian early-warning radar satellite capabilities, and removing W-88 warheads.

The United States and Russia could also begin consultations or negotiations of varying degrees of formality covering a broad range of nuclear safety issues. The most important issues, in our view, are further steps to improve Russia's access to early-warning information and steps to reduce the launch readiness of nuclear forces. As noted in the discussion of Option 7 (see Chapter Four), reaching agreement on launch readiness reduction will be extremely complex and difficult. Any progress in this area is likely to be made in small steps, perhaps starting with silo-based missiles. Russia, because of its current dependence on land-based missiles, is very sensitive to any steps it believes will make its forces more vulnerable to attack. This does not mean that launch readiness reduction is impossible, but merely that a good deal of mutual trust will be needed to make it a reality, particularly for far-reaching proposals that affect all nuclear forces. The

United States and Russia should begin discussions on measures to reduce launch readiness at this time.

Another near-term step is that of beginning discussions on how to implement a destruct-after-launch (DAL) system. We recommend that a midcourse DAL system capable of disabling the reentry vehicles be considered. Although such a system is more complicated than a boost-phase system, it provides an additional 15 to 20 minutes of decision time, compared to the latter's 3 to 5 minutes. We are aware, however, that once discussions begin, the technical difficulties of setting up a DAL system (for example, reengineering reentry vehicles and changing the yield on nuclear weapons) may prove difficult to overcome.

MEDIUM-TERM STEPS

Within five to seven years (the medium term), additional steps to improve nuclear safety may be possible. By this time, the difficult negotiations on reduced launch readiness may begin to bear fruit. We believe that simple, straightforward steps are needed, ones that can then be expanded, if and when appropriate, to include more forces and more-difficult platforms. It is hoped that the U.S. willingness to immediately reduce the alert rate of 150 missiles as an immediate step will serve as an important impetus for getting this process started. To this end, the first step we recommend here is for Russia and the United States to place an equal number of silos (around one-third of the U.S ICBM fleet, or 150 silos) off alert. For the United States, these will be in addition to the 150 silos it took off alert in the first phase, at the start of its efforts to improve nuclear safety. The sensors used to monitor the silos would be slightly modified versions of the early-warning sensors placed on the silos earlier (see Option 2 in Chapter Four). The second area in which consultations may have borne fruit by this time is a consensus that a DAL system will (or will not) improve nuclear safety and perhaps some agreement on the best way for each country to proceed.

During this medium-term period, the United States might also take two unilateral actions to continue improving nuclear safety. First, if the United States and Russia have achieved a greater level of trust, the United States may be able to adopt a new deterrence posture, one that involves moving away from a nuclear posture based on rapid counterforce attacks against Russia and placing greater emphasis on a nuclear doctrine that is more flexible and not as time sensitive. Second, by the 2007–2009 timeframe, the United States may be ready to begin deploying a limited national missile defense—a step that could improve nuclear safety, but only if it were effective against a limited number of Russian missiles and did not adversely affect relations with Russia.

POSSIBLE INTERMEDIATE- TO LONG-TERM STEPS

Implementation of the options will be affected by the direction of U.S.-Russian relations and other geostrategic issues that might affect the two nations' nuclear postures. Given the uncertainty, we cannot recommend specific intermediate- to long-term steps (next 10 to 20 years) for improving nuclear safety. However, to reach an end state in which launch readiness is significantly reduced, a series of steps will have to be taken. Figure 5.3 provides a general framework for nuclear safety steps that could be implemented during the second decade of the 21st century. The important point is that any steps taken in the

10 to 15 years	End States
Possible Intermediate to Long-Term Steps	**Possible End States**
• Reduce number of SSBNs at sea	Extensive monitoring significantly reduces launch readiness of nuclear forces in all nuclear states
• Provide global EW	OR
• Build joint U.S.-Russian missile defense	Some U.S. and Russian nuclear forces remain on modified alert; others are taken off alert with modest monitoring; and political relations between U.S. and Russia are similar to those between Britain and France (i.e., allies with no concerns about nuclear forces and no need to verify alert status)
• Begin negotiations on verification measures for reducing launch readiness of submarines, mobile missiles, and bombers	
2018	End State

NOTE: SSBN = ballistic missile submarine; EW = early warning; DAL = destruct after launch.

Figure 5.3—Potential Steps for Improving Nuclear Safety, Beyond 2010

next decade will have to build on progress made during the current decade.

As noted previously, negotiations to reduce launch readiness are likely to continue for an extended period of time. If silo monitoring and reduction of launch readiness are successful, Russia and the Unites States might contemplate including other systems, such as mobile missiles and submarines, in future plans. To further induce Russia to agree to additional reductions in launch readiness, the United States could trim the overall size of its at-the-ready deterrent by reducing the number of Trident submarines it keeps at sea.

Two other joint steps are possible in the intermediate to long term. First, if the United States and Russia prove successful at working together on early-warning issues, they could consider setting up a global early-warning system. The United States and Russia might be able to jointly provide information to an early-warning center in any country that wants it. This could be particularly helpful in South Asia, where China, India, and Pakistan lack any real early-warning system and where the potential for accidental or unauthorized use of nuclear weapons is high. The United States also might consider working with Russia on a joint missile defense system.

POSSIBLE END STATES

If nuclear safety continues to be an important objective over the long term, two end points are possible for U.S. and Russian nuclear forces. The first is a world in which all nuclear powers keep their nuclear forces at extremely reduced levels of launch readiness. In this world, all nuclear forces would be extensively monitored to ensure that no nation could quickly raise the alert level of its forces and thereby gain an advantage. Obviously, this world would require that the United States and Russia reach agreement with all of the other nuclear powers, which is why it seems unlikely to come about.

The second possibility is a world in which the United States and Russia keep a small number of their forces on a modified level of alert and their remaining forces off alert. This arrangement would serve as a minimal deterrent to attack by another nuclear-armed state. However, this world requires that U.S.-Russian relations resemble those of Britain and France today. In other words, neither country would be concerned about the size and posture of the other's forces, and no monitoring or verification would be required. Like France and Britain, the United States and Russia would each be capable of launching a nuclear attack on the other, but for all practical purposes such an attack would be unthinkable. In terms of nuclear safety, the danger posed by the U.S. and Russian forces would be greatly reduced from what it is today.

CONCLUSION

The phased approach to the Nuclear Safety Initiative that we recommend here is based on the premise that nuclear safety, U.S.-Russian relations, and U.S. security more broadly are inextricably linked. Progress in one area will improve the situation in another. Given the improving relations between Russia and the United States and the emerging security context for the United States, there is now a historic opportunity to address one of the more vexing problems left from the Cold War: how to reduce the risk of accidental or unauthorized nuclear use to as close to zero as possible.

President Bush has signaled a strong desire to pursue a "new strategic framework" that will remove nuclear weapons as part of the equation for U.S.-Russian relations. With a series of bold unilateral moves, he could demonstrate his seriousness and, at the same time, improve nuclear safety by reducing Russian anxiety about U.S. capabilities and intentions. These moves could have a positive effect on U.S.-Russian relations, one that could set the path to further improve nuclear safety and to enhance U.S. security in other important areas, such as nonproliferation and counterterrorism.

Nuclear safety cannot be improved, however, without a sustained, coordinated effort. That effort must start with Presidential commitment and leadership. Moreover, the complexities involved suggest that the U.S. and Russian militaries will have to work closely together to achieve success.

ILLUSTRATIVE PRESIDENTIAL DIRECTIVE FOR IMPROVING NUCLEAR SAFETY

One of the difficulties in designing a set of strategies and policies for improving nuclear safety is the broad, cross-cutting nature of the problem. Nuclear safety touches on pivotal and controversial issues—for example, nuclear strategy, the readiness and posture of U.S. and Russian nuclear forces, the changing nature of U.S.-Russian relations—so any effort to design useful strategies and policies is likely to be both technically challenging and politically sensitive. These difficulties are likely to be compounded by the need to integrate any steps to improve nuclear safety with other U.S. and Russian goals and policies. If past is prologue, it is safe to say that no nuclear safety measures will be implemented without first facing a variety of bureaucratic obstacles, in both governments.

Within the U.S. government, the President would have to articulate the imperative to improve nuclear safety. One mechanism he could use to do this is a Presidential Directive. In it he would task the departments and agencies to define potential options for improving nuclear safety, to evaluate them against a set of specific criteria, and to make specific recommendations. The President would then make a decision based on the recommendations. This Appendix presents an illustrative example of a Presidential Directive on nuclear safety.

Of course, the President could instead decide to take another route, one in which he and a few close advisors formulate the policy and analyze the options. This is the approach that President Bush's father used to develop unilateral cuts in 1991, when he was President. Both approaches have their advantages and disadvantages. But both require Presidential leadership and strong commitment for success.

ILLUSTRATIVE PRESIDENTIAL DIRECTIVE ON NUCLEAR SAFETY

Background

The reactions of both the United States and Russia to the attacks of September 11, 2002, have solidified the evolutionary changes that have been occurring since the end of the Cold War. For the first time since the Second World War, the United States and Russia find themselves cooperating, in this case in the war on terrorism. Despite these seismic shifts, however, substantial elements of the strategic nuclear forces of both nations remain on alert. This posture endangers both nations because of the possibility it provides for an unauthorized launch by a terrorist or an accidental launch based on mistaken information. These grave dangers are exacerbated by Russia's economic difficulties and by the large number of U.S. counterforce strategic nuclear weapons that are ready to strike within minutes.

In light of these concerns, the President has decided to make two objectives—reducing nuclear danger and improving nuclear safety—a priority of his administration. The United States urgently needs to work with Russia to improve the nuclear safety of both nations.

Goals

Relevant Federal agencies are directed to explore technical, operational, policy, and diplomatic measures to meet the following three goals. Moreover, since the promotion of nuclear safety will involve many dimensions, all three should be pursued as part of a comprehensive strategy.

1. Ensure that both the United States and Russia have reliable, high-quality early-warning and attack-detection capabilities.

2. Extend the time that civilian and military officials have to make decisions involving the possible use of nuclear weapons both in peacetime and during crises.

3. Reduce the risk of accidental and unauthorized use of nuclear weapons that could arise because of

- An unauthorized but intentional launch by a terrorist or rogue commander.

- A training accident.

- The potential misinterpretation of a benign event (space launch, sun glint, etc.).

- The misinterpretation of a nuclear event—i.e.,

  - A nuclear attack by a third country or terrorists.

  - An accidental nuclear detonation.

These goals should be pursued as part of an overall U.S. policy of improving U.S.-Russian relations and redefining U.S. deterrence needs in light of a rapidly evolving geostrategic environment. The process of improving nuclear safety should involve immediate and near-term steps to build confidence and trust between the two countries, thereby enabling more-extensive steps in the medium and long term.

Nuclear Safety Options

The President directs the Secretary of Defense to take the lead in defining and analyzing a series of options for improving nuclear safety. Because of the complex operational issues involved, close consultation with the military will be essential. Options should be defined for each of the goals above, building on past initiatives and our experience in dealing with Russia. In consultation with the Secretary of State and the Director of the Central Intelligence, each of the options should also be analyzed in terms of whether they would best be undertaken unilaterally by the United States, mutually with Russia through informal agreements, or through formal negotiations and treaties. These options should include, but not be limited to, the following:

Goal 1: Ensure that both the United States and Russia have reliable, high-quality early-warning and attack-detection capabilities, paying particular attention to filling gaps in Russia's early-warning network.

- Improve arrangements to share early-warning information either bilaterally or multilaterally.

- Provide funding and/or technology for construction of Russian early-warning radars or construction or launch of satellites.

- Establish a joint, redundant system for warning of ICBM attack by placing sensors near each other's ICBM silos.

Goal 2: Extend the time that civilian and military officials have to make decisions involving the possible use of nuclear weapons both in peacetime and during crises.

- Immediately stand down all nuclear forces to be eliminated under proposals offered at Crawford, Texas.

- Pull all ballistic missile submarines out of range of Russian targets.

- Reduce day-to-day launch readiness by one or more of the following (or other) methods:

  — For ICBMs, by disabling silo doors and removing guidance systems or warheads.

  — For ballistic missile submarines, by removing launch-critical components or keeping fewer at sea.

- Increase the survivability of some nuclear forces and command and control systems.

Goal 3: Reduce the risk of accidental and unauthorized use.

- Share command and control technology and personnel reliability procedures.

- Install post-launch destruct mechanisms on strategic missiles.

- Deploy limited missile defenses of the United States.

Evaluation of Options

Each option should be evaluated using the following criteria:

- Contribution to reducing the risk of nuclear use.

- Effect on current U.S. strategies and targeting plans.

- Effect on U.S.-Russian political relations.

- Effect on other major international actors (China, Europe, etc.).

- Effect on prospects for achieving nonproliferation and counter-terrorism goals.

- Feasibility and affordability.

- Effect on incentive to strike first with nuclear weapons.

- Ability to monitor or verify implementation of the option and the effect of cheating.

In coming to an overall assessment of each option, a negative evaluation based solely on one of the criteria should not automatically mean rejection of that option.

Recommendations and Timeline

Improving nuclear safety will be a long-term process with immediate, near-term, and medium-term measures implemented in a phased manner. An interagency working group will be convened to review the Department of Defense's initial set of options with the goal of designing a timetable for possible implementation. Options will be grouped both by how rapidly they can be implemented and by whether they can be accomplished unilaterally, through rapid mutual agreement, or only through extensive technical negotiations.

For options requiring the active support of the Russian government, the President, the Department of State, and the Department of Defense will need to engage their Russian counterparts, the goal being to mutually design a series of steps to improve nuclear safety and to then take the necessary implementing measures. Military-to-military consultations will also be important for addressing the complex operational issues likely to arise.

The end state sought by the United States is a significant improvement in nuclear safety along with a strengthened U.S.-Russian relationship.

BIBLIOGRAPHY

Arbatov, Alexei, Vladimir Belous, Alexander Pikaev, and Vladimir Baranovsky, *De-alerting Russian-U.S. Nuclear Forces: The Path to Lowering the Nuclear Threat* (Moscow: Institute of Global Economic and International Relations), October 2001.

Bailey, Kathleen C., and Franklin D. Barish, "De-alerting of U.S. Nuclear Forces: A Critical Appraisal," *Comparative Strategy*, Vol. 18, January-March 1999, pp. 1–12 (available at http://nipp.org/5.php).

Ballistic Missile Defense Organization, *Harnessing the Power of Technology: The Road to Ballistic Missile Defense from 1983–2007*, September 2000, http://www.acq.osd.mil/bmdo/bmdolink/pdf/power.pdf.

Ballistic Missile Defense Organization, "National Missile Defense Integrated Flight Test Three (IFT-3)," BMDO Fact Sheet 110-00-11, November 2000.

Blair, Bruce G., *The Logic of Accidental Nuclear War* (Washington, DC: Brookings Institution) 1993.

Blair, Bruce G., *Global Zero Alert for Nuclear Forces* (Washington, DC: Brookings Institution), 1995.

Blair, Bruce, statement before House National Security subcommittee, March 13, 1997.

Blair, Bruce G., "Dealerting Strategic Nuclear Forces," Chapter 6 in Harold A. Feiveson (ed.), *The Nuclear Turning Point: A Blueprint for Deep Cuts and De-alerting of Nuclear Weapons* (Washington, DC: Brookings Institution), 1999, pp. 101–127.

Blair, Bruce, "Russian Roulette," interview for *Frontline*, aired February 23, 1999 (available at http://www.n/wgbh/pages/frontline/shows/russia/interviews/blair.html).

Blair, Bruce, Harold Feiveson, and Frank von Hippel, "Taking Nuclear Weapons off Hair-Trigger Alert," *Scientific American*, November 1997, pp. 74–81.

Congressional Budget Office, *The START Treaty and Beyond* (Washington, DC: CBO), October 1991.

Downing, John, "Russian SSBN Patrols Halted for Three Months," *Defense Week*, January 11, 1999, p. 1.

Edenburn, Michael W., Lawrence C. Trost, Leonard W. Connell, and Stanley K. Fraley, *De-alerting Stragetic Ballistic Missiles* (Albuquerque, NM: Sandia National Laboratories), Cooperative Monitoring Center Occasional Paper/9, SAND 98-0505/9, March 1999.

Forden, Geoffrey, "Letter to the Honorable Tom Daschle Regarding Improving Russia's Access to Early-Warning Information" (Washington, DC: Congressional Budget Office), September 3, 1998.

Forden, Geoffrey, "Letter to the Honorable Tom Daschle on Further Options to Improve Russia's Access to Early Warning Information" (Washington, DC: Congressional Budget Office), August 24, 1999.

Forden, Geoffrey, *Reducing a Common Danger: Improving Russia's Early Warning System* (Washington, DC: CATO Institute), May 2001.

Frankel, Sherman, "Aborting Unauthorized Launches of Nuclear-Armed Ballistic Missiles Through Post-Launch Destruction," *Science and Global Security*, Vol. 2, No. 1, November 1990, pp. 1–20.

Garwin, Richard, "De-alerting of Nuclear Retaliatory Forces," Amaldi Conference, Paris, France, November 20–22, 1997.

Garwin, Richard L., "Post-START: What Do We Want? What Can We Achieve?" testimony to the U.S. Senate Committee on Foreign Relations, February 27, 1992.

Hoffman, David, "I had a Funny Feeling in My Gut," *Washington Post*, February 10, 1999, p. A19.

International Foundation for the Survival and Development of Humanity, *Reducing the Dangers of Accidental and Unauthorized Nuclear Launch and Terrorist Attack: Alternatives to a Ballistic Missile Defense System* (San Francisco, CA: International Foundation for the Survival and Development of Humanity), January 1990.

Johnson, Celeste, "Estimating Costs and Technical Characteristics of Selected National Missile Defense Systems" (Washington, DC: Congressional Budget Office), January 2002.

Karas, Thomas H., *De-alerting and De-activating Strategic Nuclear Weapons* (Albuquerque, NM: Sandia National Laboratories), SAND2001-0835, April 2001.

Lewis, George N., and Theodore A. Postol, "The Capabilities of Trident Against Russian Silo-Based Missiles: Implications for START III and Beyond," presentation at "The Future of Russian-U.S. Strategic Arms Reductions: START III and Beyond," a meeting held in Cambridge, MA: February 2–6, 1998.

Miasnikov, Evgeny, Center for Arms Control, Energy, and Environmental Studies, Moscow Institute of Physics and Technology, Moscow, interview on January 28, 1999.

Missile Defense Agency, "Missile Defense Milestones (1944–2000)," http://www.acq.osd.mil/bmdo/bmdolink/html/milstone.html.

Molander, Roger, David Mosher, and Lowell Schwartz, *Nuclear Weapons and the Future of Strategic Warfare* (Santa Monica, CA: RAND), MR-1420-OSD, 2002 (limited distribution; not for public release).

National Institute for Public Policy, *De-alerting Proposals for Stragetic Nuclear Forces: A Critical Analysis* (Fairfax, VA: National Institute for Public Policy), June 23, 1999.

Norris, Robert S., and William M. Arkin, "Nuclear Notebook: Russian Nuclear Forces, 2001," *Bulletin of the Atomic Scientists*, Vol. 57, No. 3, May/June 2001, pp. 78–79.

Nunn, Sam, "Amendment No. 4180 to the FY97 Defense Authorization Act," Section 1303, National Missile Defense Policy.

Podvig, Paul, "The Operational Status of the Russian Space-Based Early Warning System," *Science and Global Security*, Vol. 4, 1994, pp. 363–384.

Podvig, Pavel (ed.), *Russian Strategic Nuclear Forces* (Cambridge, MA: MIT Press), 2001.

Postol, Theodore A., "The Nuclear Danger from Shortfalls in the Capabilities of Russian Early Warning Satellites: A Common Russian-U.S. Interest for Security Cooperation," presentation to Carnegie Endowment for International Peace, February 26, 1999.

Rumsfeld, Donald R., *Foreword to Nuclear Posture Review Report* (Washington, DC: Department of Defense), January 8, 2002, p. 2.

Sessler, Andrew, et al., *Countermeasures: A Technical Evaluation of the Operational Effectiveness of the Planned U.S. National Missile Defense System* (Cambridge, MA: Union of Concerned Scientists and MIT Security Studies Program), April 2000.

U.S. Department of State, *The Treaty Between the United States of America and the Union Of Soviet Socialist Republics on the Reduction and Limitation of Strategic Offensive Arms (START)*, http://www.state.gov/www/global/arms/starthtm/start/toc.html.

Velikhov, Evgeny, Nikolai Ponomarev-Stepnoi, and Vladimir Sukhoruchkin, *Mutual Remote Monitoring* (Moscow: The Kurchatov Institute), January 26, 1999.

"The Watchers Fall Asleep in Orbit," *Obshchaya Gazeta*, No. 20, May 2001.

Weldon, Curt, "Defense of America's Homeland," speech delivered on the floor of U.S. House of Representatives, May 2, 2001 (available at www.house.gov/curtweldon/speechmay22001defense.htm).

William J. Casey Institute of the Center for Security Policy, "As Expected, Russia Gets a Bail-Out—But It Won't Get Moscow Through New Year, or Protect U.S. Security Interests," *Perspective*, No. 98-C 128, July 1998 (available at http://www.security-policy.org/papers/1998/98-C128.htm).

Wolfstal, Jon, "Kursk: Cold War Causality," *Christian Science Monitor*, August 28, 2000.